LIVING IN BRAZIL

As a Peace Corps Volunteer and Businessman

H. LYNN BECK

LIVING IN BRAZIL
AS A PEACE CORPS VOLUNTEER AND BUSINESSMAN

Copyright © 2018 H. Lynn Beck.

Cover Image Taken by: Julia Goughnour

All rights reserved. No part of this book may be used or reproduced by any means, graphic, electronic, or mechanical, including photocopying, recording, taping or by any information storage retrieval system without the written permission of the author except in the case of brief quotations embodied in critical articles and reviews.

iUniverse books may be ordered through booksellers or by contacting:

iUniverse
1663 Liberty Drive
Bloomington, IN 47403
www.iuniverse.com
1-800-Authors (1-800-288-4677)

Because of the dynamic nature of the Internet, any web addresses or links contained in this book may have changed since publication and may no longer be valid. The views expressed in this work are solely those of the author and do not necessarily reflect the views of the publisher, and the publisher hereby disclaims any responsibility for them.

Any people depicted in stock imagery provided by Getty Images are models, and such images are being used for illustrative purposes only.
Certain stock imagery © Getty Images.

ISBN: 978-1-5320-6136-3 (sc)
ISBN: 978-1-5320-6135-6 (e)

Library of Congress Control Number: 2018913601

Print information available on the last page.

iUniverse rev. date: 11/15/2018

Contents

Dedication ix
Acknowledgment xi

Peace Corps Training 1
First Assignment: Cuiabá, Mato Grosso 5
A Trip North of Cuiabá, Deep into the Forest 11
Moving to Natal, Rio Grande do Norte 20
David 26
Portuguese 29
Green Beans, Cassava, and Sun Meat 30
Making Pizzas 32
Alecrim on Market Day 33
Riding the Buses—Snapshots into People's Lives 35
Trip Back to the States to Visit PhD Programs 37
Learning to Dance the Samba in Brazil 39
Carnival 45
A Nightclub in João Pessoa 48
The Engagement 49
The Yacht Trip 52
The Marriage 55
Back to the USA—First Time 57
Back to Brazil 63
Trip Home to See Dad 69
Milk Study 73
Becoming Sick 77
Part-Time Consulting 79
David versus Goliath 84

The House of Representatives and the Men's Club 86
Running on the Street next to the Beach 87
Nicholas Becomes Ill 88
Christmas 90
Barbecue on the Corner 91
Gas Stations Closed for the Weekend 92
Trip to Paraiba for Consulting 94
Consulting for the Algodoeira 97
Opening a Computer Store 99
Tarantula Mating Season 102
Large Rats Invade Our House 103
Voodoo versus White Table Spiritism 105
Getting Things Done in Northeast Brazil 108
How to Counteract a Macumba Spell 113
On Using Candy for Money at the Supermarket 117
Another White Table Visit 119
Garbage on Our Lot 120
My Neighbor with a Machine Gun 122
Employee Problems 124
Chased by a Motor Scooter 125
Sued by Rio Grande do Norte's Attorney General 127
My Friend Killed Outside a Nightclub 129
Move to São Paulo 131
Finding a House in São Paulo 133
Two Weeks at a Convention in Rio de Janeiro 138
Leaving the Company 140
The Chicken Ranch 143
The German Pig Farmer 147
Surprise—All Prices Are Frozen 151
Winter in São Paulo 155
The Police and the Thief 156
Our Worker and the Street Thieves 158
Going to the Bank on Payday 160
Driving the Beltway 161
Banco Safra 162

Preparing to Return to the US 168
Home Again after Eight Years..170
Finally, a Job 171
1988 and Beyond 175

Dedication

I am forever grateful for my best friend, Dona Katia. I am also grateful for her family: Dona Vania, Dona Naide, and Seu José. Without their friendship, I never could have survived ten years in Brazil.

I am grateful for the support my children: Kevin, Nicholas, and Christianne. Their encouragement kept me writing.

Acknowledgment

Illustrations by Michael Ries

Cover Photo: Michael Maxey, Dick Goughnour, Harold Lynn Beck

PEACE CORPS TRAINING

I finished my master's degree in Vermont in mid-1974. I had no idea what I should do, and when in doubt, join the Peace Corps. I filled out an application, and eventually, I received an invitation to work in Brazil. I accepted. I felt that it was a perfect job. My assignment had me working in education in the state of Mato Grosso. An advantage of going to Brazil was that I would learn to speak Portuguese and become familiar with a major culture in Latin America. I began counting the days to the start of training.

All trainees were shipped to Philadelphia for processing, after which we were bussed to New York, where we flew from New York to Rio de Janeiro. It was a very long flight, but all the trainees were too excited to sleep. Everyone spent the night jabbering about their personal lives. I was excited too, but I kept to myself.

From Rio we flew to Belo Horizonte, the capital of the state of Minas Gerais

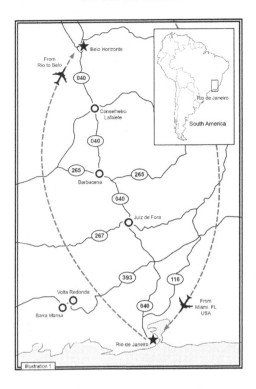

Illustration 1

Belo Horizonte was a very large city. Minas Gerais, which means "general mines" in English, was known for its mining of emeralds, rubies, diamonds, and other precious stones.

The male trainees were split between two or three boarding houses, as were the female trainees. I was placed, along with five or six others, in an old lady's multistoried rooming house. It was difficult to learn Portuguese while mixed in with several English-speaking trainees.

Each day we had to take a bus to the Peace Corps training center. Some of the other trainees stuck close to me because they knew that I spoke Spanish, and could solve any problems that might arise on the way.

At the training center, the main building was a house located on one side of the property. Several rooms located around the outer wall of the property were used for individual language classes. There was a center courtyard used for meetings, games of volleyball, and occasional drinking moments.

It was nothing like my first Peace Corps training experience, when we had trained on the mountaintop in a rain forest in Puerto Rico. I soon learned that the trainees were very different as well. Most had joined the Peace Corps as a means of enriching their résumés. Helping people was secondary to their goal of improving their résumés. I was not impressed by most of them. I had an even more difficult time trying to relate to them than I normally did with people. I kept to myself. The Peace Corps had changed since my first experience in 1967, and I did not like this experience as much as the old one.

We had three to seven trainees per language instructor, but I quickly found myself unhappy. I felt that I could learn much faster than the other trainees because of my fluency in Spanish. I became frustrated. Soon, I stopped going to class and stayed in the main building, reading books in one corner of the library.

Word circulated that I was not attending class, and soon I was visited by the person responsible for our Portuguese training. After I presented my case, one staff member mentioned that he had a friend who was a surveyor in the rural area. He suggested that I could live with his friend's family and follow him around. He called his friend Victor, who agreed to accept me. The next day, I was off to Victor's house, via a bus from Belo Horizonte to Victor's town. I had not felt comfortable living in that huge city. It had made me nervous.

Victor met me at the bus station, which consisted of the bus parked under a large shade tree on the town square. Victor was personable, and he hustled my bag into his Jeep and drove me to his home. He chatted as he drove. I understood half of what he said, but my Portuguese did not allow me to uphold my end of the conversation. As he carried my bag inside his small house and set it beside the couch, he mentioned that his wife was at work.

Victor explained that he had to survey a ranch and we would be in the field for the rest of the afternoon. He drove to a small outdoor snack bar, and we ordered a couple of sodas and ham-and-cheese sandwiches. Before we departed, he told me he had a partner, João, whom he had to pick up at his house a couple of blocks over. João was already standing by the street. As soon as we stopped, he jumped into the back seat, and we were off.

My Portuguese consisted of 95 percent Spanish and 5 percent Portuguese, but we were able to communicate. Thanks to my previous Peace Corps experience, I was relaxed at being on my own with limited language ability. I always found a way to communicate. If my Portuguese and Spanish failed me, I still had hand signals and the English–Portuguese/Portuguese–English dictionary.

We started on a two-lane paved highway with broad shoulders and no potholes. After a few miles we turned onto a less-traveled side road. Again and again, we turned onto less-traveled side roads until we were on a one-lane dirt path that was filled with dips and holes and that passed around shrubs and over cattle gates. Suddenly, Victor pulled over and parked. He explained that this ranch was owned jointly by two brothers, but since they both had married and started their own families, they needed to separate the land and building assets into two equivalent ranches. That was Victor's job.

For me, this was very boring. Victor set up his instrument, checked for levelness, and after sending João walking away from us with his survey pole, started taking distance and angle readings. It was very hot and dry. I was sweating profusely and soon became thirsty. Victor did not seem fazed by the heat or the sun. I saw no beads of sweat on him, whereas sweat was streaming down my face. I left him to his work and did not try to talk with him. I did not want him to regret his decision to allow me into his family's life.

It took all afternoon to finish the job. I was glad to see João returning, dragging his surveyor's stick. I could see that he was tired. Victor loosened the screws on his instrument and repacked it in its box, and we retraced our path home.

I lived with Victor and his family for several weeks. I knew it was not always comfortable for them to have me in their small house, but they never allowed their frustration to show. I was paying them rent for the use of their house, but I think they accepted me into their home not for the rent, but to do a favor for their friend who had asked on my behalf.

My Portuguese improved a little each day while I was there. I always tried to learn a new word each day, but with my limited vocabulary, it was possible for me to learn a half dozen new words each day.

Our training period ended, and I was called back to the training center. Peace Corps had its swearing-in ceremony, where we were sworn in as volunteers. Afterward, a few very cold beers were consumed by all, and then we were sent to our final destinations: our places of work.

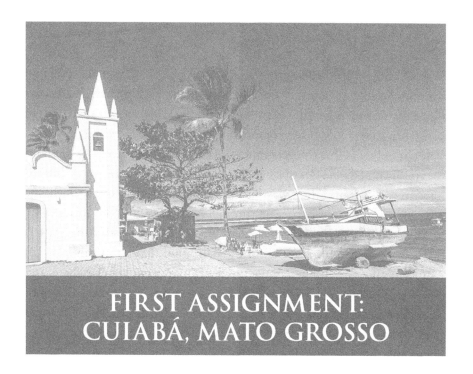

FIRST ASSIGNMENT: CUIABÁ, MATO GROSSO

I was sent to Cuiabá, Mato Grosso—the geographic center of South America

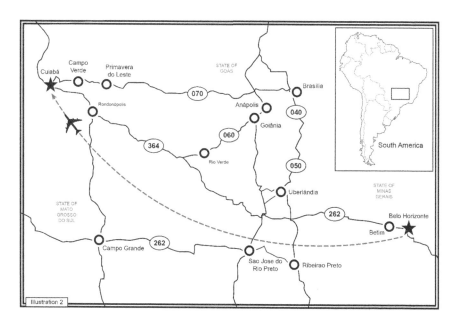

When I stepped off the plane in Cuiabá, I felt like I had been dumped into a pressure cooker. It was very hot and humid, and we were in the early part of the dry season, not in the rainy season. In the rainy season, it would rain constantly and would be even more humid.

Not only was this the geographical center of South America, but also the continental divide ran through Cuiabá. To the north, the water drained into rivers that found their way into the Amazon River. To the south, the water drained into rivers passing through Argentina and into the ocean.

An implication of being the geographic center of South America was that Cuiabá was located farther from civilization than any other place in Latin America. Anything manufactured was manufactured elsewhere, and elsewhere was always half a world away. Everything had to be shipped in from the industrial south: São Paulo and its surrounding region. The roads for most of the distance between São Paulo and Cuiabá were poor, almost impassable during the rainy season and filled with potholes during the dry season, making transportation costs high. The cost of living in Cuiabá was the most expensive I had seen anywhere. The only product that was inexpensive was lumber. There was no shortage of lumber because trees were being cut everywhere to clear land.

Farming in southern Brazil was very advanced. After World War II many Germans had decided to move to Brazil and had settled in the southern states. There were many cities where people everywhere spoke German; even schools were taught in German, until the federal government passed a law requiring all schools and government business to be conducted in Portuguese. The architecture of most buildings in many cities was predominantly German.

In the south, the demand for land was high because everyone wanted to farm and own his own land, but landowners not only did not want to sell any land; they also wanted to buy more land. When the government opened rural Mato Grosso for development, there was a land rush. People aspiring to own large farms, such as small farmers and hired men, rushed north to grab as large a piece of land as was possible. If they had land, they rented it to a neighbor and left their families while they went north to locate suitable land and clear it for farming. Only then would they bring their families north.

This surge north had consequences. The city of Cuiabá and the surrounding region were growing very fast. The population now was predominately male. Housing was scarce and very expensive. Jobs were difficult to find and low-paying. There was much hustle and bustle. Hardware stores were selling axes, spades, chains, chainsaws, and nails. Everyone had a backpack or a mule to carry his provisions into the wilderness.

When I arrived, I was told that there were some problems with my assignment and that they needed to be resolved before I could start. Until then, I would be in a holding pattern.

The Peace Corps director told me to find a place to live and wait. This was not easy because our living allowance was minimal relative to the escalating costs of living. I was told of a bunkhouse-style boarding house located on the edge of the city. It consisted of a large room with no room divisions. On each side were rows of narrow beds. There were about thirty or forty beds in all, and rarely were any vacant. In one corner were shower and bathroom stalls. Males rented the beds by the day, week, or month. It was not a secure place, and nothing of value could be left there, including while you slept. I left my billfold inside my pillowcase. At night, it was hot beyond comprehension, there was no ventilation, and mosquitoes were a major problem. Everyone who could afford one bought a rotating fan and arranged it carefully to blow slightly above their body, sweeping from head to foot. This minimized the risk of catching a cold in the heat yet dissuaded the mosquitoes from landing on our bodies.

I met some men at the bunkhouse who had come up from Rio Grande do Sul: a great agricultural state far to the south. They all envisioned that one day they would own large farms that they could bequeath to their children. They spoke very confidently, as if it were a fact that simply had not happened yet. These men were preparing to disappear into the forests and stake their claim, and then they would fell huge trees using their axes and chainsaws. These determined men would try to burn the fallen trees as quickly as possible and throw seeds into the soil, expecting that crops would jump out. They were often disappointed. Forest soils were good for trees, but not so much for crops.

Many of the trees being cut were mahogany. The size of a tree was determined by a measurement where men faced the tree, stretched out their hands, and held the hands of the men standing next to them. The measure was how many men it took to encompass the girth of the tree. It was common for a tree to require three or more men to embrace it. The trees were burned where they fell. They had no value.

The lack of reasonable roads, even in the dry season, prevented the trees from being transported for processing into planks. Besides, no sawmills capable of handling the large trees were available. Even if the wood could have been processed into planks, it still would have had to travel thousands of miles to find any market; therefore, the trees had no value. Cuiabá was located at the end of the world—where the wind goes before it stops to rest and turn around. It was like Cuiabá was an island located in the middle of the Pacific Ocean.

Many of the residents from the bunkhouse were friendly. We often sat on our beds and talked about our families and why we were in Mato Grosso. It was from these citizens of Rio Grande do Sul, also known as Gauchos, that I learned of *erva matte*, or green tea. I learned that it was part of an exquisite social ceremony shared in the most basic situations, but always with great significance. The tea was carried in a bag as one carried tobacco for smoking. The container in which it was placed for drinking was called a *cuia*, from the name of the gourd from which it was made.

The host, the man who suggested that we partake of the *erva*, continued the conversation as he carefully loaded the *cuia* with tea. The *cuia* was always five or six inches deep and two or three inches wide. It was filled to within one inch of the brim and then tipped on its side to prepare an open space going from the top of the *cuia* to its base. Then the man produced a long silver straw with a quarter-sized filter at the bottom. He placed the metal straw carefully along the side until it reached the bottom of the *cuia*. He slowly righted the *cuia* and reached for his thermos bottle, which was always filled with recently boiled water. He poured it into the *cuia* along the straw. The first suck from the straw was the host's. This was because it contained small pieces of leaves he deemed too unpleasant for his guests. After the host was confident that the tea was no longer filled with leaf fragments, he passed it to the next person, who sipped and passed it to the next and the next, until it was gone. Then the process would start anew. It was wonderful.

It was at this point that I became aware of the boil on my neck. It was very painful and grew worse each day instead of better. It was especially painful in the hot climate, and my shirt constantly irritated it by rubbing it. The Peace Corps authorized me to have it removed surgically. I had to find my own way to the hospital, which was not easy. I did not have the money to take a taxi. I either had to walk in the heat and sweat like a fool or learn the bus routes. I walked.

Early the next morning, they put me under general anesthesia and removed the boil. I was lonely when I slowly came out of anesthesia. No one visited me, and no one would pick me up. I felt bad from the aftereffects of the surgery, simple as it had been. My family at home had no idea. It would have been nice to have a friend or family member there to talk to me.

It would be a couple more hours before I could be released. Forced immobility and not feeling well were a recipe for loneliness. I slept as much as possible and had a dream while the effects of the anesthesia were wearing thin. I dreamed two lovely ladies were walking down the corridor, and upon seeing this young man in this huge room all alone, they stopped, looked, and entered my room. They approached my bed. I believe one was the mother and one was the daughter. They were both beautiful, with bright smiles and huge eyes. One asked, "Are you an American?"

I think I smiled and replied, "Yes. How did you know?"

The other asked, "Are you here alone?"

"No. You are here with me."

This caused them both to smile even wider. One asked, "Don't you have any friends?"

Understanding that I might be able to gain some sympathy, I said, "No. I know no one." Perhaps I was pressing my luck, I thought, and they might discover my theatrical ploy.

"Oh, that is sad," said one of the ladies. They were a team and could conduct excellent conversation as they alternated questions and answers.

"Miss," I managed to say weakly.

"Yes," answered the daughter as she came close to hear better.

"Will you marry me?" I pleaded, using my most desperate voice.

I believe I saw the daughter break into a wonderful smile, as did the mother while she grabbed her daughter's shoulder and hurried her out of the room. They looked back at me as they left the room and smiled. I had not yet mastered my technique with the ladies, but I knew I was improving.

Day after day, I had nothing to do, especially since I had only enough money to pay for my boarding house, laundry, and basic food. I did manage a couple of Pepsis each night as entertainment, but that was the extent of my crazy money.

For security reasons, I left my suitcases filled with my belongings in the Peace Corps director's office. I used a small duffel bag to take a change of clothes to my bunkhouse each day and then returned with my dirty clothes to the office the next day and exchanged them for clean clothes from my suitcase. Then I went to the center of town to watch the movement, and movement there was. It reminded me of what a boomtown might have looked like during periods of rapid expansion in the old West. People were everywhere, always busy, doing something or going somewhere and buying things and taking them somewhere. I just stood and watched and waited. I was frustrated. I was envious of all those who had a purpose in their lives.

Once while I was at a bar in the center of the city, drinking my Pepsi, another American came in, ordered a beer, and sat at a neighboring table. He was about my age, but he had a long, uneven beard and wore a crumpled hat. I started a conversation with him, and he joined me at my table. He was an anthropologist hired by the government organization tasked with protecting the indigenous communities. He was an official Indian negotiator. When trouble broke out between the ranchers and Indians, he was called in to negotiate a settlement and avoid casualties.

He told me stories of Indians shooting arrows at slow-moving trains passing through their territories. He said that sometimes ranchers would invade the Indians' land, and the Indians would catch them there and pin them down in an untenable situation. Other times the ranchers would catch the Indians in a weak position and pin them down. Someone always ran to find the anthropologist, who hurried to the spot to start negotiations. He said he was busier than I might think.

We met each night and shared stories. It was relaxing and good for me to have someone to speak English with. Then one night, an Indian came running around the corner and was relieved when he saw my friend. He ran over to our table, and they spoke quickly. My friend said he had to run, and he and the Indian ran away into the night.

A TRIP NORTH OF CUIABÁ, DEEP INTO THE FOREST

The Regional Peace Corps director arranged for me to spend a few days with an experienced volunteer, Mike. Mike was twenty-two or twenty-three years old and had married a young, beautiful local girl. The volunteer was in his third year and spoke fluent Portuguese. I envied his language ability. He worked in agriculture, supervising a nursery that produced grafted fruit trees to be given to the pioneer farmers.

Mike confided in me that he was afraid to return to the United States. He had been immersed in Brazilian culture for so long that he was not sure how he would adapt back to the American culture. He was concerned about how his wife would adjust since she did not speak any English and she was very attached to her family. She had seen her mother every day of her life. He told me that he was thinking of going to his state's university and majoring in horticulture and then hurrying back to Mato Grosso to get a job doing what he had been doing.

Mike's Peace Corps duties included supervising many workers, some of whom were agronomists, yet Mike had never gone to a university. This was often the case. Peace Corps volunteers often had responsibilities far beyond what they could obtain in the United States. For me, Peace Corps offered four advantages: first, helping people; second, learning to speak fluently a foreign language; third, learning how to live in and understand a culture different from your own; and fourth, gaining responsibility far above what you would see for one or two decades in the US.

Mike's young wife prepared a very tasty meal: black beans, rice, and a small piece of meat accompanied by some local cheese. It was simple but perfect. We had a nice conversation, always in Portuguese, since Mike had difficulty with English and his wife spoke no English. He spoke it so seldom that he thought and dreamed in Portuguese. We retired to our rooms to sleep early because we had to be up early to travel the next day. I rested on the sofa, although I could not sleep. I stood and looked out the window. The darkness was complete. There were no streetlights—none. I could not distinguish any objects outside the window. It was absolute darkness.

Early the next morning, we drank strong coffee and ate French bread. We quietly entered the Jeep and drove to an open gas station. It was still totally dark outside. We filled five cans with gasoline, five gallons each, and situated them in the back. This was an old-style Jeep with a canvas top, similar to those used in World War II. Mike pulled a blanket from the Jeep, gave it to me, and ordered me to completely soak it in water. After I soaked it in a tank, I handed it to him. With no explanation, he folded it several times and tucked it tightly around all of the gasoline cans from bottom to top.

And then we were off into the dark, north of Cuiabá

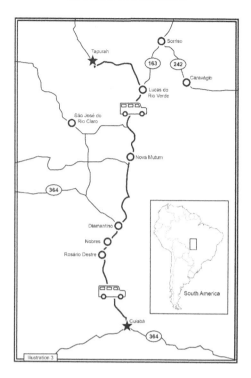

Mike lived in Diamantino ("small diamond"), which was a couple hours north of Cuiabá, and we were headed north again from there.

He drove and drove, and eventually, the sun came up. The road was no wider than the Jeep and was full of ruts and large pools of water from the recent rains. His ability to maintain his speed under those conditions was possible only if he knew the route well—and he did.

Wilderness had been around us only a short time before, but now it was almost gone. It too was moving north as the pioneers continued their land clearing. On both sides of the road, we saw huge stumps and the remains of unburned trees intermixed with green corn one or two feet tall, growing among the tree remains. As we continued north, the corn became smaller, and the tree remains were larger and larger. Finally, there was no corn, and the trees had been felled but had not been burned yet. Somehow, the farmers had missed the opportunity to burn, and the rains had arrived, washing the unprotected dark earth away in great volumes. Gullies of various sizes were growing visibly larger each day. My heart ached as I watched the conservation nightmare unfold mile after mile. It was a disaster.

Then I smelled and saw smoke but saw no fires. As we proceeded ever farther north, the smoke became denser and stronger in smell. My eyes were watering, and Mike weaved around invisible dangers in the road. It was obvious that he knew every spot in the road. Then I saw the fire. It was on both sides of the road, burning through huge trees. I could see the fire and hear the crackling.

Mike pulled to the edge of the road and carefully tugged at the corners of the wet blanket to ensure that all the spare gas tanks were covered. He asked me to watch that the blanket did not leave any part of any gas tank exposed. I understood and kept a close eye on the tanks. There were sparks blowing in the breeze and falling on the Jeep. It was hot, and our windows were open, as was the back of the Jeep. Sparks were everywhere. Occasionally, one would escape my eye and land on my pants or sleeve, drawing my attention only when I felt the burning sensation on my leg or arm. This was frightening and painful, and we were driving into it. After a few miles, the fires stopped, and the smoke cleared. We had made it through the fire alley.

After another half hour, we were still continuing north, but suddenly, the vehicle lost power and rolled to a stop. We were caught in the middle of a huge pool of water created by the rut that was the road. The road was just a bit wider than the vehicle. We stepped from the vehicle without getting our feet wet because the ground was the same level as the Jeep's floorboards.

I knew nothing of the mechanical mysteries of vehicles. I thought we were doomed. I asked how far we were from a gas station or a mechanic, or from anywhere. "About fifty miles," Mike said, but he showed no concern. I was frantic.

Mike stepped onto the front bumper and threw open the hood, which now rested on the windshield. He checked a couple thingamabobs and ordered me to open the cubbyhole to retrieve a screwdriver, which I immediately did. Mike was the most serious Peace Corps volunteer I had ever encountered. He never wasted a word. I watched as he removed the ignition cap and found a doohickey and separated it from the other thingamabobs. He announced, "This is the problem. It's broken."

I remained unconvinced that the problem was so simple. I asked, "How long will it take us to return to civilization?"

He pulled a huge knife from somewhere and began whittling on something while he asked, "Why?"

I said, "Well, if it's broken, do we not have to go buy a part to repair it?"

He said, "Let's walk up the road. There is a house, and we can see if they have any old radio batteries." Then he answered my question. "No, we don't. I can make the part we need."

I kept my mouth shut and tried to keep up because he already was yards ahead of me. On the left we saw a huge pasture and, next to it, a large plantation of cassava. In one corner of the cassava plantation, there was a home garden and next to it was an old shack. The shack had a grass roof and dirt floor, and small tree branches had been used to make the walls. Like most rural houses, it had no door. The house was well ventilated. In the field several people, each with a hat, were bent over hoeing a crop. They were all sizes, genders, and ages.

Mike announced that we would try them. As we approached, he explained to me that they were immigrants from Japan and had accumulated a few thousand acres of land, of which they cleared a little each year. They produced everything they needed and had purchased only a radio and batteries, because there was no electricity within forty miles. He also said that they had two thousand head of cattle, and the land and animal assets they owned far surpassed a million dollars in value, yet from age three to one hundred, they worked twelve hours a day, wore homemade sandals, and lived in a hut with a dirt floor and see-through walls. They lived poor in spite of being wealthy. This was the dream that all people from the south had when they migrated to the north. This was everyone's dream.

Mike led me to the oldest gentleman, where he inquired whether the family had any old batteries. The other family members did not stop to watch or listen. They continued their task, which required them to keep their eyes on their plants to avoid mowing them down along with the weeds that surrounded them. Unfortunately, they had no batteries. Mike graciously thanked the man, and we immediately returned to the rutted road.

Mike never slowed his pace. He said, "There is another house a mile or two down the road." And we continued to walk at his fast pace.

Soon, a hut appeared a short way from the road. I followed him to the house. The walls were made from bamboo, with ample cracks for air circulation. There was a doorway but no door. I was concerned about how he would knock. He walked to within ten feet of the house's entrance and stopped. He clapped his hands and yelled, "*Oh de casa!*"

I learned that *Oh de casa* was a polite way of knocking on a nonexistent door. It was not a good idea to walk up to a house without a door and knock on whatever was available to knock on. The house owner might take offense at your looking into his home without his permission.

After a second the owner appeared in the doorway and smiled. He took the two steps down, since the house was built two feet off the ground, and shook our hands. He noticed that I continued to stand a step behind Mike. He told me not to worry as he waved his hand in the direction of the steps into his house. I did not know what he was talking about. That was when I saw a small bamboo cage located at the side of his steps and, in it, a sizable snake looking straight at me. I jumped back. That was when he really laughed. He was enjoying himself. He told me that the snake was only ten to twelve feet long and was still a baby. He used him for rat control. I noticed that he also did not have any dogs or cats as pets.

Mike quickly turned the conversation to what mattered. The farmer did have an old battery and was happy to donate it to our good cause. The gods were smiling because he had just replaced his radio's batteries and had not yet disposed of the old ones. There were two, and Mike asked if he could use both of them. Mike offered thanks, and we departed. There were no wasted motions by either the farmer or Mike.

We found the Jeep exactly as we had left it. Mike grabbed the old part and looked at it. It was a piece of graphite. He explained that batteries also had a graphite bar inside them. He took out his super knife, opened the battery, and removed the graphite, which he compared to the failed part. He whittled and compared and whittled and compared. Within a few minutes he tried the newly fashioned part. It fit. He replaced all the thingamabobs, closed the hood, and ordered me inside. I was there instantly. He turned the key. The engine fired, and we continued our travels.

Farther down the road, we stopped to visit another Peace Corps volunteer. This PCV was living near a new city that was only six years old but already had 27,000 inhabitants. We found the PCV in his yard, and we all talked a little. This PCV did not seem to be hyper or particularly busy. He and Mike informed me that this region was a new frontier, and no one living in the region now had been here seven years ago. Everyone had come from somewhere else and for varied reasons, but they all had one reason in common: they appreciated the fact that no law and order existed in the region. In the frontier towns, there was no law; therefore, a certain class of citizens favored living there. Everyone was armed with a pistol or knife and knew how to use both—and probably had used both.

Both PCVs counseled me to avoid card games and drinking in groups, especially with people I did not know well, and to never inquire into the history of any citizen. People did not take kindly to being asked where they came from. If I asked the question, they might suspect me of representing law and order. They preferred the city without law and order since many of the citizens had, at some time, run afoul of it. People had died or disappeared because a local citizen mistook another citizen for a person representing the law.

All three of us crowded into the Jeep and entered this new city. I was fascinated. The houses were all constructed from mahogany planks, mostly unpainted. The floor on each house was raised a couple of feet from the ground, and the space under the floor was surrounded by a lattice of narrow wood. The wall planks were always separated by half an inch. This might have helped ventilate the houses. The roofs were made from tile. The city of 27,000 did not have a paved or cobblestoned street. All of the roads were dirt. Along the sides of the streets were vehicles parked in all ways, and a few horses were tied to rails placed next to the roads for that purpose.

My companions selected a restaurant, and we entered. The floor also was constructed of wooden planks with half-inch spaces between them. We sat at a table, and a waitress came and took our orders. The other PCVs were acquainted with several people and had cross-table conversations. I noticed that the men they spoke with wore cowboy hats and had pistols strapped to their waists.

When the meal arrived, we were quick to start eating. In my excitement, I dropped my knife, and it fell through the crack in the floor. I knelt on the floor and lowered my head to see if I could retrieve the knife. Mike yelled at me not to stick anything between the cracks. I looked down, and my eyes widened as I distinctly saw two eyes looking back at me. The guys ordered me back into my chair and quickly fetched the waitress to replace my knife.

I saw from my companions' facial expressions that they had underestimated my stupidity. They explained that the people placed anaconda snakes under the houses to control pests. The snakes were limited in movement to the space under the house by the lattices that surrounded the space between the ground and the floor. This arrangement was convenient because it allowed rats to the area under the houses. The snakes were content to stay under their respective houses because their food came to them and they were always well-fed.

About this time, a waitress with some free time was sweeping the floor. She did not need a dustpan because she guided things into the space between the boards. Sometimes small pieces of food were dumped there. It was explained that the snakes liked only live food, but the bits of food that were swept between the boards attracted rats, which kept the snakes busy.

We returned the local PCV to his house and chatted more there. He and Mike were remembering their early years as volunteers. Mike told of a trip he had made with another PCV in a dugout canoe on a huge nearby river. As Mike paddled, he kept the canoe away from the edges. The new PCV asked him why he did this. Mike said that when he was recently arrived, he had paddled near the shore and under overhead branches. As he was passing under a branch, an anaconda had dropped down and bitten his thigh. Anacondas' bites were not venomous, but they hurt and could become infected. The snakes did this to secure their prey while they dropped down and coiled around them. Since there were two men in the canoe and they both had knives, they were able to ward off the snake. Since then, Mike had always avoided the river's edges.

On another day, the river was very hot and humid, and paddling had made the same new PCV hot. On an impulse, he had asked, "Hey, can we stop for a little and swim?"

Mike said no.

The new PCV asked, "But why not? The river is clear. There is no danger. I only have to be careful when I dive to avoid those logs that are deep in the water."

Mike replied, "Those are not logs," and continued paddling.

After returning Mike's friend to his home, we returned to Diamantino. I thanked Mike for his kindness and caught a bus back to Cuiabá. I went to the Peace Corps office for an update on my job. It was not good. The job had disappeared. The Peace Corps director was arranging for me to be transferred to Natal, Rio Grande do Norte: a city located on the beaches in sunny Northeast Brazil. Even better, my director said that within three months he would be transferring to become the regional Peace Corps director in Recife, a city four hours south of Natal. I liked him and knew that I could trust him. I was excited.

MOVING TO NATAL, RIO GRANDE DO NORTE

After thirty days in Cuiabá, my job had disintegrated. I had been informed that I would have to accept the position in Natal, Rio Grande do Norte, or return home. I was not ready to return home. I looked at Natal on the map. It was very far away, but it was indeed on the beach. I thought it might require some adjustments from me since Natal had more than a half-million people, but I was a flexible person.

Our regional Peace Corps office was in Recife, Pernambuco, a city of one million people and also located on the beach. I flew into Recife from Cuiabá and received my orientation at the Peace Corps office

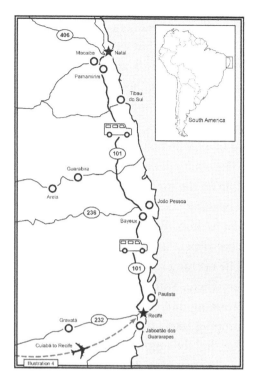

It was a huge city, but I liked it. Once all the bureaucratic problems were resolved, I was transported five hours to the north along the coast and introduced to my supervising agency. I would be working in the equivalent of our state agricultural extension office.

At the time, I did not know what I was supposed to do, but now, forty-five years later, I understand perfectly: I was expected to do nothing. They expected nothing from me. They wanted nothing from me. They hoped that I would be quiet, invisible, and undemanding and that I eventually would go away. They had so many major problems to resolve that my idiosyncrasies were not high on their list of priorities. They were kind, and they gave me a desk and a chair and asked nothing from me. In exchange, I had a reason to live in Brazil for two years, to learn the language and the culture and maybe even mature more. Looking back, I had a wonderful deal.

I had to find temporary living quarters, so I rented a room in a cheap hotel in the downtown area until I could manage to locate a permanent place. I had to find a place rapidly since my living allowance was being consumed by the hotel room rates. The cost of living in Natal was much lower than it had been in Cuiabá, but we received a smaller living allowance.

Since I knew no one, I walked around the downtown region alone. The streets were so busy, so full of people. Residents were walking up and down the street, and if they bumped into me, they never said they were sorry. When I bumped into them, they were surprised when I apologized. Some even stopped to see who had spoken to them. They knew I was from way out of town. The wide sidewalks were filled with people. In addition, people were standing on the street selling hot dogs, candy, other food, and lottery tickets and begging. Each stationary person interrupted the flow of people and caused a backup, like a wrecked car on an interstate highway.

I entered a large department store. The stores there were different from our stores. The opening to the street ran the full width of the store. Displays were everywhere, with little room for people to walk. People were constantly bumping into each other, with no one taking offense or even noticing. Bumping into other people was as normal as breathing. Customers must be slightly aggressive, or they would go home without purchases. Merchandise was often knocked to the floor and trampled on, but no one noticed.

It was very warm, and I was sweating profusely after walking only a few steps. After walking several blocks, I returned to my hotel room at noon and took a shower, but it did no good. I was sweating again within minutes. Not only was it hot; it was humid.

I had my clothes crunched up in my suitcases and needed to find someone to wash them. Clothes were always washed by hand. I had to count each item in each laundry batch to be sure that everything was returned. It took a couple of days, but my clothes were always returned fresh and perfectly ironed.

During this period I was living mostly on cookies. My living allowance was not large, and if I had eaten in a lunch place, especially all three meals, I would have had no money left after the first week in the month. I went to supermarkets to study their inventory and prices. I noticed how dirty their floors were compared to our supermarkets. They had so many fresh fruits and vegetables. Grapes fell to the floor, and people stepped on them, leaving stains and wet spots. No one was concerned about the untidy floor or that someone might fall. In Brazil customers did not sue businesses for anything. Customers knew the condition of the floors and took preventive action by sidestepping. The floors were periodically cleaned, just not as often as in the US.

I soon became acquainted with other Peace Corps volunteers living in the city. There were three who were finishing their term and would be leaving in a couple of weeks. They lived in adjoining houses on the beach. The houses were small, maybe four hundred square feet, and old, but they were no more than a few steps away from the beach, and each volunteer could afford his or her own house, if the volunteer lived prudently. I was doing my best to adapt to all the sacrifices required to be a PCV. If it would allow me to live in my own house on the beach, I would make the sacrifice.

On my first day on the job, I was taken around and introduced to the other people in the office building, which was a former two-story house. Rooms were small, and most could barely accommodate two desks. Larger rooms could accommodate three or four small desks. The walls were bare in the room to which I was assigned, except for an outdated and crooked calendar. The man with whom I would share the space was not there my first day. He appeared on the second day. He was an American from Kearney, Nebraska, a farm boy who had been brought up using scrapers to level land for irrigation. He was about sixty-five years old. He had married a local lady, and they had adopted a girl. He had fought in World War II and had been a POW for several years in Poland. My upbringing was similar to his. I too was a farm boy who had been brought up using scrapers to level land for irrigation. We had much in common.

Not knowing what to do with my time, I tried to read Portuguese books, but I found my language ability still severely lacking. I carried on anyway, since my goal was to learn a new word each day.

There were two new volunteers trying to rent the beach houses as the older PCVs vacated them. I quickly visited with the third volunteer and received her blessing to rent her house after she left.

Finally, the day came, and I moved into my new house. I was so happy. I had to buy a mattress, which I threw on the floor. I bought a fan that had a sweeping motion to help me combat the mosquitoes and stay cool at night. I bought all the kitchen pans and equipment in the house from the exiting PCV. I walked to the supermarket and bought bread, cheese, ham, eggs, rice, and beans—all the essentials to survive.

I grew to dislike my lack of work at the state extension service. It was a long walk from my house on the beach to work, maybe two miles or more. With the heat and humidity, I was always sweaty and miserable. On the plus side, I had many good talks with Emil, my American neighbor at work. The Brazilian employees did not seem busy, nor did they seem concerned that they were not busy. They spent most of their time drinking coffee, reading newspapers, and chatting. They would come late to work and leave early. No one cared.

I always hurried home as soon as I could in the afternoon because I could capture some time on the beach. I enjoyed playing Frisbee with myself. When there was a strong sea breeze, I could throw the Frisbee into the breeze, and the breeze would bring it back to me. For more exercise, I could throw it at an angle and then run like the devil to catch it. It was like throwing myself a pass in football.

There were few people on the beach at that time, but a handful of people always congregated to watch me play Frisbee. It was beautiful—the sun was setting, people were sitting together on the beach with the breeze blowing their hair, and the bars were turning on their lights and preparing the tables for their evening guests. That was my signal to head home.

My house was located on a small rock outcropping on the beach. The outcropping was wide enough to maintain a street down the center, no more than two or three hundred feet long, with houses on both sides. My friend Dick had a small house on the street. On one side of his house was another house, and on the other was a corridor used by people to reach the beach. It was only six feet wide. This was the same corridor that passed by my house and separated my house from the wall of the nearby bar. This bar stretched from the street to the beach. There was a small covered area by the street, but most of the bar was outside. There were coconut trees growing there, surrounded by outdoor tables that were always occupied. The spot offered a gorgeous view of the beach and the city, although guests sat under the coconut trees at their own risk.

Next to Dick's house and heading into the corridor were Mike's house and then my house. My house consisted of two bedrooms, each big enough for a small bed and dresser. It also had a small kitchen and living room. The living room was wide enough and long enough to hang a couple of hammocks. I used it for a small picnic table, a hammock, and a couple of easy chairs. Behind the kitchen was an open area where the maid did the laundry. There sat a tank, three by four feet and at least three feet deep, that I used to bathe at night under the stars. I had a *cuia*, a half of a gourd, that I used to pour water onto my body. At night, it was cold, but on a hot day, it was refreshing.

On Saturday mornings, around nine or ten o'clock, Dick, Mike, and I, along with any other volunteers visiting from other rural sites, would meet at a bar just across the street on the highway that followed the beach. It was already hot, and a really cold beer tasted sweet.

The bar's lot was deep and went up a very steep hill. Only the first thirty feet from the street were useful for a business. After that, the hill's slope was too steep. Both sides of the bar consisted of vacant lots filled with weeds, small and large. The bar's owners built a four-foot-high adobe wall to surround their property. As the waiter brought each beer, he would remove the cap and toss it over the wall into the neighbor's lot. Each time he did it, he scared the rats. They would scatter in all directions. We loved watching the weeds part as they ran through them. We always speculated about how large they must have been.

By eleven o'clock the beaches started filling up, and by one o'clock they were full of people. By three o'clock only a few people remained. That was when I liked to go to the beach. Once, I was walking along the beach, minding my own business, when a bunch of young ladies lying on the beach started whistling at me and calling me over to them. They scared me, and I ran away. I was not used to that. Brazilian ladies were very forward, and I had not yet become accustomed to that.

I was to learn that the entire northeastern region of Brazil had more women than men, considerably more. This was caused by the periodic droughts that destroyed agriculture and livestock production and related employment. This lack of production fed into other industries and led to further unemployment. To find employment, the working-age men moved to São Paulo to find jobs and never returned. Women could not migrate because they had to stay behind with their families. This resulted in many more women in the Northeast than men.

DAVID

David was a force of nature. Everyone in the state of Rio Grande do Norte knew David or wanted to know David. He was famous, and everyone loved him. When I had met him during my training in Belo Horizonte, he had presented himself as a quiet, timid man. He had come to our training, given his serious and succinct presentation on horticulture, and left. In Rio Grande do Norte, his image was different.

David was from a New York farm. He had graduated from Cornell University with a master's degree in horticulture and had been one of the first people to join the Peace Corps after its formation. He had been sent to Rio Grande do Norte, and after he finished his two-year tour, he had stayed in-country with only a few coins rattling in his pockets. He had begun working for several agro-related companies. For one job, he traveled the countryside buying corn from farmers, which was to be used by Purina to manufacture animal feed. He became intimately acquainted with the geography of the state and discovered the best place to farm. He soon had found the perfect little farm with an absentee owner; first, he rented it, and later, he bought it. It was next to the Açu River and three hours' travel by car from Natal. Until David arrived, no one had considered that the river water could be used for irrigation.

David started producing vegetables, and as he produced them, he developed different technological packages for each vegetable. He could not go to the extension service for help because—and I am ashamed to say so—they did not know how to give that help. The technical advice given by the extension system was for states located far away from Rio Grande do Norte, which had very different soils and climate. David was innovative, optimistic, and self-reliant. He did not attempt any project before he knew it would work. He conducted small test trials to find the optimal technological package for each crop he wanted to produce before he applied it to his farm on a larger scale.

Once he had mastered the technological packages, he sought special markets to obtain the highest price. He found that the off-shore oil platforms located off the coast of Natal were perfect. They demanded high-quality food and would pay for it. David could produce it. He secured the contracts and went about producing. All he needed to do was buy an old VW van to make the three-hour trip from his farm to the point of delivery. This was not easy since he had no capital to start his farm. He continued working his jobs during the day and worked the farm at night and on weekends.

David lived in a small, old, three-sided machine shed, where he slept in a hammock slung from two posts. He bathed in the river and ate his meals in bars in town. The town was a half mile from his farm. He ate mostly rice and beans, chased down with ice-cold beer. David loved his beer. His work uniform consisted of a pair of shorts, a light shirt, a pair of sandals, and a floppy hat to keep the sun off his neck and face. His needs were minimal. David was a complex man, but one with simple needs.

He himself grabbed the oxen and ran them up and down the fields to make plow furrows that could be used for planting. He threw the harness over his shoulder and grabbed onto the handles attached to the long, sharpened log used to make the furrow. The scene was reminiscent of mid-nineteenth-century Nebraska.

There were days when I heard a knock at my door and was surprised upon opening it to see David. The first thing he asked for was an ice-cold beer. Nothing tasted worse than a moderately cold beer. Brazil's beer always tasted good if it was served ice-cold. My refrigerator was too old and broken-down to keep beer really cold; however, I did have a bar next door. All I had to do was open my door, take two steps until I was facing the six-foot-plus wall surrounding the bar, and yell, "*Oh, José, passe duas geladinhas por cima!*" I would hear José grunt a confirmation, and within a minute, two beers would appear on a tray over the wall. I took the beer, placed the money on the tray, and held the tray above the wall until José secured it. David and I could then get down to business.

David always talked about the farm—his crops or his small machinery or new things he was thinking about doing. He always had a new market he wanted to produce for. He dreamed of sending his produce straight to Paris, France. I enjoyed listening to him. David had no concept of time. He might arrive at any time and stay for hours. Indeed, there were times I had to cast him into the night because I needed to sleep to be productive the next day. David could drink until 3:00 a.m. in Natal, make the three-hour drive to his farm, hitch the oxen, and spend all day in the hot sun, making furrows. David had a source of energy that I envied.

One Sunday David knocked loudly on my door. As soon as he knocked, I knew who it was. He knocked aggressively and with purpose. When I opened the door, there was David, except he was wearing a pair of old 1960s-style glasses with thick black plastic rims. These were secured by a string tied around his neck and attached to the nose part of the glasses frame. He had on a T-shirt, a pair of swimming shorts, and a silly smile. I also noticed some seaweed hanging from his ear and his glasses. He entered my house and asked for a beer. I stepped outside and yelled for José, and two cold beers appeared on top of the wall.

David told me that he had been partying on the other side of the river with friends, but they had wanted to take a nap after drinking and eating. David had wanted action, so he had swum across the river where it flowed into the ocean. Depending on what part of the cycle the tide was in, this could be very dangerous due to the undertow. David did not seem to mind. David drank a few beers and then disappeared as quickly as he had appeared. He apparently had other places to go.

PORTUGUESE

Each month my Portuguese became better. We three PCVs spoke English among ourselves, but not all the time. We all had girlfriends and had to speak Portuguese when they were present. The best way to learn another language was to find a girlfriend or a boyfriend. If you were married or were a couple of English speakers, your chances of improving your Portuguese went to almost zero. For guys, the key was to find a local girlfriend, and then you would be speaking Portuguese pronto, lickety-split. When the heart has something to say, the mouth must find a way!

GREEN BEANS, CASSAVA, AND SUN MEAT

On a Saturday, we decided to track down and eat at a restaurant famous on a national level. It was listed in most good travel books as a "must experience" attraction. Its specialty was green beans, cassava, and sun meat. The first item was green beans. The individual beans were removed from the pod while they were still soft and then boiled with butter. The second item, cassava, was a root crop used like Americans used potatoes. It had a long, fibrous root that was peeled, cut into smaller pieces, and boiled. Instead of gravy, Brazilians used butter for flavoring. The last item was the main course, like turkey at Thanksgiving, except that for you to understand better, I must provide some background.

On ranches in rural Brazil, there was rarely any electricity; therefore, there was no refrigeration. If an animal was butchered, the meat would quickly spoil if there was not some special way to prepare it. Over the centuries, the people had learned to cut the meat in thick slices and soak it in special condiments, including salt mixed in milk. After the meat absorbed the salt and condiments, they threw the meat onto wire fences during the day. The hot sun had its way with the meat and drove the salt deep into its center. I have no idea what else they did to the meat, but it would no longer spoil without refrigeration.

The restaurant was in the middle of a very poor district located near the beach. The streets were typically constructed from cobblestone, and lack of maintenance allowed gigantic potholes to form. It was almost impossible for a car to reach the restaurant. There were no signs to help us find it. We would walk a block or two and then ask some resident for directions. Sometimes the locals did not even know what we were talking about. The restaurant had more fame across Brazil than it did for the people living one block away.

When we found the restaurant, it had no signs or other indications that it was a restaurant. We found it because its doors and windows were open, and we smelled the food. It was just one of the houses in the community, except that it consisted of one large room filled with homemade wooden tables and, instead of chairs, little stools. They were not even painted.

We were lucky to find an unoccupied table. We sat. There was no ordering. They had one plate: green beans, cassava, and sun meat, accompanied by very cold beer. I have never tasted any better combination of food and drink. The salty sun meat called for a sip of cold beer, and a sip of cold beer required a bite of sun meat. I have eaten in many places in my life, but none rivaled that food on that day. It was for that reason that people were known to come from all over Brazil to sit on those stools and eat that food. I felt blessed that we could walk to such a place.

MAKING PIZZAS

I found an old recipe book lying in a corner of the house, left by the previous PCV, and within it I found a recipe for pizza. I decided to give it a try and made a trip to the supermarket. The sacks of ingredients were heavy to carry home since they included flour, a small ham, and tomato paste and sauce, but I powered through the walk. When I reached my house, I needed a cold beer to start, so I did the "one cold one over the wall" routine, and José obliged.

I struggled with the dough, but I blindly followed the directions. While I waited for the dough to rise, I mixed the sauce ingredients and cut the ham into small pieces. The ham was the most expensive ingredient and nearly broke my bank. It put me in emergency mode for the rest of the month. When the dough was right, I stretched it into the pan and spread the sauce over it, and into the oven it went.

My first pizza was made in secret. I did not want to have guests over and have nothing to feed them if my experiment failed, but it did not fail. When I took the pizza out of the oven, it was beautiful and exuding wonderful smells. I ran to see if Dick and Mike were home. I wanted to share my good fortune and a cold beer with them.

I tried to eliminate these paragraph things, but I could not.

ALECRIM ON MARKET DAY

Alecrim was an old town that Natal had grown up around and absorbed. It was at least three hundred years old. Its streets were all narrow, constructed from uneven cobblestone that required people to look down as they walked to avoid stumbling. The sidewalks were narrow, often insufficiently wide for a person to walk. The houses were narrow as well and shared walls on the sides. Most houses were just wider than was needed to stretch a hammock, but they went deep into the block. On any day of the week, the streets were alive with people, but one day a week, the streets were jammed: that was market day in Alecrim.

In larger cities, each little town inside the city had its own market day. Each market had its own specialties, such as fabrics or cooking utensils or spices, but all markets had the basics. I loved going to Alecrim on its market day, especially if I was feeling lonely or depressed. The colors of Alecrim were astonishing. So many people walked through the streets, and each was dressed in bright colors. In addition to the people, there was the merchandise: clothing, fabrics, plastic materials, fruits, vegetables, and spices. It was a color festival that would elevate anyone's spirit.

Each store spread its wares outside its narrow store space. Those wares tumbled onto the sidewalk and even spilled into the street itself. There was a sound fest as well. The store workers yelled and waved their arms, trying to attract the attention of people walking by. The scraping of cart wheels and the clank of shoed horse feet against the cobblestones interrupted the screaming of street hawkers.

People walking in the streets constantly bumped into other people while trying to find what they were looking for. They had to dodge other people doing the same. Many porters were scurrying back and forth while carrying large loads on their backs. Other porters were bringing merchandise to the stores to be sold yet that day. It was chaos. I would usually buy some mangoes or papaya or cashew fruits or pineapples to take home with me. The fruits were all soft from ripeness and emitting their fragrances. I never wanted to go home, but my legs and arms eventually tired, and I had to find a bus.

RIDING THE BUSES—SNAPSHOTS INTO PEOPLE'S LIVES

Natal was founded on December 25, 1599. Natal, in English, means Christmas. It was built on the beach and against a river. It grew out from that starting point. After a short distance of lowlands, the geography changed, and elevation increased. In fact, the little outlet on which I lived was part of the lowlands. As soon as we crossed the coastal highway, there was a steep increase in elevation, perhaps as much as a hundred feet in a horizontal distance of five hundred feet.

The lower city, the old city, was called Ribeira. It had been built in the area wedged between the upper city, the river, and the ocean. It was even older than Alecrim. Its streets were narrower, more crooked, and more uneven. The houses were narrower and longer, with more people everywhere. Every time I took a bus from my house to the upper city, I passed through this neighborhood.

When I took a bus during the evening, I always felt like I was at the movies as I looked out the window. This was because the fathers were home playing with their children in the front rooms of the houses while the mothers were in the next room, the kitchen, preparing the evening meal. One particular scene has always remained in my mind. The father was seated on the sofa, holding the hands of his daughter, who could not have been more than two years old. She was dancing, and the father was encouraging her by moving their clasped hands to the beat of the music and singing. The mother had her face turned toward the two and was smiling. That was how Brazilian women learned to dance so well—they started as soon as they could stand up. They were encouraged by their fathers and mothers and practiced daily.

The scene that I remember so well could not have lasted more than half a second because the bus was moving at a good speed. But most of the houses had their front doors open, and I caught a split-second view into each family's life. It was like watching a moving, where each home was a frame as I sped past, catching but a glimpse into their lives.

TRIP BACK TO THE STATES TO VISIT PHD PROGRAMS

I had eight months left on my second Peace Corps enlistment when I decided I wanted to study for my doctorate in agricultural economics. I felt that I needed to visit some programs before I made my choice. I applied to Peace Corps for special permission to return to the US, at my expense, to visit a few key universities. The organization gave me permission.

I flew to Sacramento, California, to view the university there. I felt no connection after visiting with a couple of professors. I flew home to Nebraska, where I visited University of Nebraska–Lincoln. I also visited the Universities of Missouri, Kentucky, Tennessee, and Iowa. They were all excellent universities with admirable programs, but I felt a connection with the University of Missouri. The university offered me an assistantship on the spot, which I accepted gratefully.

I did not then know that this was going to be a difficult road. I had been accepted into a PhD program in agricultural economics. I had only basic mathematic and economic skills, no statistics experience, and no ability to write. On the PhD level, it was imperative to possess excellent English grammar. I did not. While working on my PhD, I would have to take remedial classes in calculus, micro and macroeconomics, statistics, and writing. Had I fully understood what was coming, I would have asked for an assistantship for a master's degree rather than a PhD.

It was good to see my family during this visit. Dad had effectively retired and had invited Mom to learn to dance. They drove to Omaha, a hundred miles each way, to take dancing lessons twice a week. They wanted to buy me a gift, so I asked for samba lessons. I was having difficulties learning the dance by doing it in Brazil. I wanted to surprise my dates by learning the step-by-step breakdown of the samba. My parents arranged for me to have a few hours of individual instruction. I was grateful. I was confident that I would surprise my dancing dates.

LEARNING TO DANCE THE SAMBA IN BRAZIL

Upon my return, I decided to apply my new knowledge of dancing the samba. On the first Saturday night, I took a bus to the nightclub and waited for the partying to start. People didn't start to enter the club until around 11:00 p.m. The dancing was great after 1:00 a.m. I noticed a shapely blonde enter with two companions whom I deduced to be her sister and brother. After watching them for a few minutes, I was convinced I was right. I stood and approached her table. I gathered all my courage and asked her to dance. She seemed happy to accept. She had a wonderful smile that revealed her soul—it was a happy soul.

I discovered immediately that she was a motivated dancer. She placed her arms around my neck. I was not sure what to do, so I grabbed her waist. She was already into the beat of the music, and her hips were moving around like a washing machine in the middle of a spin cycle, only she was going up and down as well as around and around. I was not so much dancing as hanging on. As she tossed me around the dance floor, I tried to relax and enjoy it, but my main concern was not being thrown into other people or a wall. Her hips were moving in ways I could only imagine because I was too close to actually see what she was doing. I tried to look around to determine if anyone was laughing, but no. To other people, everything we were doing apparently seemed natural. I had no chance of showing her my moves.

After a few dances, she invited me to her table. Since my table had been taken over by other people, I decided to accept. I explained to her that I did not know how to dance but wanted to learn. I am sure that she was already aware of that, but she was not concerned. She was happy that she had an opportunity to dance and that I was capable of hanging on or following. She asked a few questions: Who was I? What was I doing in Brazil? Did I miss my family? And then we returned to the floor to dance again.

Later, we agreed to meet again the following Saturday at 11:00 p.m. And so it came to be that we met every Saturday night at 11:00 p.m. There was no way that I could apply the samba that I had learned in my private lessons in the US. Once, when I tried, my partner asked me what I was doing. I told her that I had learned to do the samba while I was in the US. She told me to stop it. She advised me to feel the music and follow the rhythm. The problem was that I had no rhythm. In the end, I clung to her waist and tried not to lose her on the dance floor. She was fun to dance with and nice to look at.

One day when the four of us (her sister and brother were always present) were sitting at the table, she started asking me personal questions. Did I have a girlfriend in the US? Did I like Brazilian women? After several more questions, I realized that she wanted to take our relationship to the next level. That was not in my game plan. She next asked what my religion was. I said that I was a *macumbeiro* (a voodoo priest). Her face tightened in surprise. She had not expected that, not from an American. Her smile became forced. She was very trusting of people, so it never occurred to her to ask me if I was telling the truth. She could never imagine otherwise. Then, stupidly, I added for emphasis, "And if you ask any more questions, I will turn you into a frog." I smiled, thinking I was funny. She did not smile. Her enthusiasm quickly diminished, and we soon parted ways.

The next week, I was at the nightclub by 10:00 p.m. and claimed a small table. When 11:00 p.m. came, I started to look for my dance partner, but she did not appear. By midnight, I wanted to dance. I could not just get up and dance without losing my table. Also, it was customary for men to carry purses, or *capangas*, in which they stored their money, photographs, car keys, cigarettes, passport, and other official documents. I had one and had become dependent on it. I could not dance with it; it was too large. I could not leave it on the table, or I would lose both the table and the *capanga*. I was stuck.

I started looking around to see if I knew anyone so that I could join their table. This way, I could leave my *capanga* on their table for them to watch, and I could find someone to dance with. I strained to survey the club for anyone I might recognize. There, across the room, was a guy I knew. He was with two girls. He was a good guy, but I did not like him much because he was always talking about women as his conquests, and he used foul language to do it. I thought that the two girls he was with were likely ladies of the night or at least had loose morals, and I only wanted to dance.

I grabbed my *capanga* and headed toward his table. To avoid meeting the ladies, I greeted the guy, asked permission to place my *capanga* on his table, nodded toward the ladies, and kept walking. I went to another side of the room and stood against the wall, trying to spot a suitable lady with whom I could dance. A few yards away, I saw a pretty girl also standing against the wall and not dancing. After a short wait, I decided it was safe to approach her. I needed to make sure that she was not accompanied. Her partner might have been taking a bathroom break. When Brazilian men took a bathroom break and returned to find someone hitting on their woman, they could become very volatile. It was always better to be extra cautious than get into a fight.

I approached the young lady and asked if she was accompanied. She was not. I asked if she would like to dance. She said no. I thought, *What are you doing in a nightclub if you do not want to dance?* I asked her why she did not want to dance. She replied that she did not know how. I had thought all Brazilian women knew how to dance. *What is this?* I wondered. I asked her again. She said she could not dance because she was working. I did not understand and insisted she try to dance. She agreed. Unfortunately, she truly did not know how to dance. That was when I started to understand what she had meant when she'd said she was "working." For you innocents, like I was, she was a lady of the night advertising her charms in a nightclub filled with drunken men.

Embarrassed, I made my way back to my friend's table with my tail between my legs. I asked for permission to join them and pulled up a chair.

They were laughing because they had seen my performance as I tried to find a dance partner. Conversation was difficult because I had not made a good impression on the two young ladies. In fact, later I learned that they considered me rude, crude, and arrogant, but they were willing to smile at me and converse some with me out of respect for social etiquette, which did not allow them to speak their mind or be disrespectful in any way.

I learned that the two girls were sisters. The younger one was still studying at the University of João Pessoa, located in the neighboring state. She was majoring in psychology and was home on vacation. All I could manage to learn was that their names were Katia and Vania, and they lived just off the town square by the city auditorium, where concerts were held. In fact, the next week there was to be a concert there by a pop star known all over Brazil. The two sisters showed no interest in further contact with me. I had really made a bad first impression.

During the next week, I pondered what I should do. I decided to venture up to their neighborhood and try to locate their house. I walked down the street at night, trying to remain in the shadows because I did not want to be seen. I looked into the open-doored houses until I spotted one of them. I wanted to go say hi, but I was very shy and returned home, angry at myself for being such a weakling.

I had an idea. I could prepare a nice pizza sauce and take it to their house. We could make the crust there, and everyone could enjoy a nice slice of pizza. For me, this was an expensive endeavor because the sauce ingredients were costly, especially the pork, but I would do it. The next day, when I left work, I swung by the supermarket, bought the ingredients, and happily lugged them home.

On Saturday, I hailed a taxi to help me carry my cooking pot and all my ingredients to Katia and Vania's house. I had the taxi drive a little past the house. Their father was rocking in a chair on the front porch, like a sentry determined to protect his daughters, and I needed to exit the taxi and collect my things and myself before facing the sentry.

Struggling with the weight in the huge pan, I approached the porch and tried to smile, though I do not think it showed. I asked if Katia and Vania were home. The man yelled for his wife to come out and take care of the inconvenience. A short, happy woman appeared while drying her hands on an apron. I tried to explain how I had met her daughters and that I had brought ingredients for a pizza, which I had hoped to make with Katia and Vania. She smiled and motioned for me to enter the house. She guided me into the kitchen area and asked me to sit down. She relieved me of the pizza ingredient pan and set it on a table.

The mother's name was Naide. I called her Dona Naide as a sign of respect. We liked each other immediately. I asked for Katia and Vania, and she informed me that regrettably, they were attending the Rita Lee concert and would be back in an hour or so.

I explained how I normally made pizza. She nodded and reached for a cutting board and a knife, both of which she pushed toward me. I started cutting the peppers, onions, garlic, pork, and other ingredients. She was busy doing something else, and then suddenly, there was a sizzle from a pan that had appeared on the stove burner. She took the sauce ingredients, pushed them around in the pan, and stirred a little before turning the stove off.

She reached for the flour, and without the benefit of any cookbook, including the one I had brought for my guidance; Dona Naide produced the most beautiful crust I had ever seen. Now we just needed the girls to bring the crust and sauce together for a few minutes in the oven.

To say that Katia and Vania were surprised to see me in their house would be an understatement. Seeing me in the kitchen and on good terms with Dona Naide was more shocking. We talked and ate the most delicious pizza I had ever tasted. Dona Naide was a master chef. I was able to break down a few barriers that my behavior in the nightclub had raised. My interest was in the younger daughter, Katia. As yet, she showed no special interest in me, but when I asked if I could visit again, Dona Naide replied that I should come anytime I wanted. I felt a little better. I had won over the mother; now I just had to worry about Katia and her father, José. Out of respect, I must call him Seu (Mr.) José.

Seu José was going to be a tough nut to crack. I learned that he was retired military. He was regimented in everything he did. He had served through the communist rebellion in the late 1950s and early 1960s, and he was not a fan of anyone wearing a beard. The sight caused him flashbacks to the revolution, especially in the case of foreigners with beards because they were the ones who had instigated the revolution. Unfortunately, I had a beard. I think Katia let this information about her father leak in one of our private conversations because she still wished that I would stop stalking her. But she could not just tell me that because her mother liked me. I had a powerful ally. There was no bond more powerful between a daughter's mother and suitor than that formed over making a pizza from raw ingredients.

I started visiting Katia two and then three times a week. We would go for short walks around a few city blocks. She lived in the downtown area, so there was always a lot to see. We also stayed at her house and talked. Dona Naide constantly served me orange juice or an avocado milkshake or cashew juice. I loved all of them, but cashew juice was the best. Katia had two older sisters. Vania was a couple of years older than Katia, and Tania was a couple of years older than Vania. Tania was a journalist and worked for the government. Vania was a teacher but worked in community development for the government. During this time Seu José remained on the periphery, acting like he was indifferent, but he was not. He heard, or was told, everything that happened or was said. As the head of the household, he was in charge. Luckily for me, Dona Naide was his trusted consultant.

CARNIVAL

Carnival was approaching. Carnival, to my mind, is not the equivalent of Mardi Gras. After I experienced my first Carnival, I realized that the Brazilians were professional partiers. The New Orleans Mardi Gras celebrators were amateurs. Carnival was a period of six days (Friday to Wednesday) of absolute debauchery. The entire country stockpiled beer to avoid running out during Carnival, but they always ran out near the end anyway. Married men could go on escapades with anyone they chose, as could women, although only a small group took advantage. You could dress anyway you wanted. If you wore a mask, then no one would know who you were, and your silly and obnoxious behavior could not be associated with you. You were free to be stupid for six days. There were many hookups and instances of micro-dating during Carnival.

Friends formed groups to celebrate Carnival together. These groups were called blocks. Everyone in the block dressed alike, for the purpose of helping each member recognize other members of the group. When Carnival was in full bloom, people were known to become intoxicated, in some cases very intoxicated. It was important that when members saw another member falling down, they give that person a shoulder to lean on. They could not leave a fallen comrade behind. If one block member lost his or her sense of direction and started to wander off to follow the beat of a different drum, another block member should catch the wanderer by the belt and redirect him or her back to the block.

To help themselves stay together, members of a block often tied a rope to form a large enough circle for all the members to hold onto the rope as they danced to the music. If any member fell by the wayside, his or her neighbors should notice and reattach the member to the rope. The block also discouraged nonmembers from attaching themselves to the rope.

There were small groups of drummers who wandered the streets, beating samba music with a rhythm that attracted people to dance. Each musical group attracted its own following as it meandered through the streets. When people tired of that music, they pulled aside to drink and wait for another group to come by.

There were trucks that had been converted into large, mobile sound systems. All the speakers were positioned on the first story. These could be six or more feet high and were located on the two sides and back of the truck. The sound they produced could be heard for many blocks, even a mile. There was a platform on top of the speakers, a second story. This was where the musicians and the control for the speakers were located. There were also very bright lights in all directions that blinked to attract people and encourage them to follow the trucks as they wandered slowly through the city.

The crowds that followed these trucks were densely packed. Everyone was drinking. People sometimes stopped briefly to drink a beer, or they might have hard liquor placed in a flask strapped around their neck, and then they would return to their jumping and screaming.

The trouble with dancing in the streets was that anyone could do it. Poor people from the barrios were there, and some were intent on robbing whatever they could. Revelers simply had to be aware. Since people spent hours jumping up and down, they minimized everything they carried with them. Usually, they carried a little money in some secure pocket and their native drink flask. That was all.

Many people "jumped" Carnival only in clubs. People used the term "jump" because that was mostly what the Carnival dance was: jumping up and down for six days. It was not for the weak. That was why our city of Natal had a warm-up in September with a long weekend. Everyone drank and danced to start to get into shape for Carnival. Then in December there was a longer period in which people started to build up their resistance for both jumping and drinking. Again, Carnival was and is not for the weak. It requires physical and mental strength to have a good Carnival.

Katia, Vania, several other people, and I formed our own block. We bought fabric and decorated it so we could easily recognize each other. Our block rested in the early afternoons at Katia's house. Dona Naide served us energy drinks made from mangoes, oranges, avocados, cashews, watermelon, cantaloupe, and a host of other fruit. I know that each day she went to the market early to stock up on fresh fruit for us. She was enjoying the experience as much as we were. These were happy moments. The group laughed and joked and talked while we consumed the fruit juices. Then in early afternoon we wandered into the streets and jumped a little to warm up for the evenings ordeal. People were everywhere, and all with happy faces—not one sourpuss in the bunch. That was Carnival: happy times.

We separated from dark until about 10:00 p.m. to try to nap and eat something. I seldom could do more than close my eyes. I had no appetite for anything other than sodas or juices. We reconvened at Katia's house at 10:00 p.m., and from there we all went to the nightclub and started our jumping, which continued until the sun came up. Exhausted, we each went to our own home and tried to nap until about noon. Napping was no easy task because Carnival never stopped. Music played constantly and everywhere, and people were always dancing and drinking and shouting. Rest was only a dream. The beat of Carnival was in the air and could not be escaped.

At sunup on Ash Wednesday, the band led everyone out of the nightclub where we were celebrating and into the dawn. We went to our cars and on to the beach, where we all dived into the waves. It was a tradition to mark the end of Carnival. I was ready for it to be over.

A NIGHTCLUB IN JOÃO PESSOA

Although Carnival ended Wednesday morning, Brazil did not start to move until the following Monday, and then just barely. Katia had returned to her university in João Pessoa, the capital city of the state of Paraiba, about two hours south of Natal. I was lost without being able to visit her.

On Friday night after Carnival, I took a bus and traveled to João Pessoa to visit her. She found a place for me to stay that night, and we spent the next day together. On Saturday night, she and I accompanied a friend of hers and the friend's fiancé to a nice nightclub, where we found a small table. Since we were two couples, one couple could maintain the table while the other couple danced. The dance floor was crowded, which had the advantage of facilitating close dancing. It was a night that I could not have imagined even in my best dreams. It was perfect, and Katia did not seem to remember our clumsy first meeting.

THE ENGAGEMENT

After knowing Katia for no more than thirty days, I proposed to her. She took several days to decide, and then we had to break her acceptance to her parents. Her mother was very happy, but Seu José would be no pushover. I was nervous and had done much rehearsing. He had made no attempt to lessen the strain in our relationship over the few weeks that I had known Katia. He had kept his distance and not joined in any of the fun we were having.

On the night of the deed, I grabbed another rocking chair that Dona Naide had inside the door, carried it out to the porch, and parked it next to Seu José's chair. I placed it parallel to his chair but at a respectful distance. I synchronized my rocking to his gait. He gave a "good evening" to me, and I returned the greeting to him. After a few minutes of silence, I just blurted my question, asking him for permission to marry Katia. He remained silent for a few minutes and continued to stare forward, as did I, although my heart was in my throat. He asked if I had asked Katia and, if so, what she had said. I told him that yes; I had asked her and that she had agreed. He asked if I had told Dona Naide. I told him that I had and that she approved. Then, he said, he also would approve. I stood, shook his hand, thanked him profusely, and then shook his hand again and ran into the house. He must have thought that Katia was marrying an idiot. The ladies were sitting close enough to the porch to have heard, but far enough away to be discreet. They were already smiling and hugging each other. What a day! Seu José continued rocking and staring straight ahead. He was adjusting.

It was late February, and I would leave for the US for graduate school in May. We had to be married before then. Her parents needed time to prepare for a wedding—to obtain a date in a church, to rent a reception hall, and to save the money to pay for it all. After much work and calling in of favors, the date was set for May 8, 1976.

To square things with the church, the priest had to collect data from us. For Katia, that was easy, but for me, not so much. The first question was difficult: what was my address? I did not know. I did not receive mail at my house. So I returned home and found someone walking on the outcropping on which my house was located. I asked what it was called. The person said Bat's Point. *Okay, that's a start*, I thought. I went to the six-foot-wide corridor that led from the center of the outcropping into the sea. I asked someone what that was, and I was told, "Corridor of the Sexual Perverts." *Okay*, I thought again. My house was the third house in the corridor. I returned to the padre and told him that I had discovered my address. He perked up, smiled, and looked at his secretary, who prepared to type my response. I announced that I lived on "Bat's Point, Corridor of the Sexual Perverts," and that I was pervert number three. The secretary tried to contain a smile and looked at the padre for guidance. He gave a holy smile and said that he would fill that in later.

I had one good friend at CEPA (in English, the State Commission for Agricultural Planning), where I now worked. He was a young economist, married with a baby. He agreed to be my best man. He invited Katia and me to his house so that I could call my family and tell them of my engagement. That would be his gift to us, since these calls were very expensive. I remember that Katia and I sat on a long sofa, with Eduardo and his wife sitting in chairs to my right. We had been served coffee and had completed the required small talk with our friends when Eduardo handed me the phone and told me to dial away. I did.

After a short wait, Mom answered. She was shocked because we had never spoken on the phone while I was in Brazil, although I had been home only six weeks before. I told Mom that I was engaged. She was speechless. I had not had a girlfriend six weeks earlier. Suddenly, I felt strange and looked to my left, where Katia used to be, and she was at the far end of the sofa, looking at me like I was from Mars. Apparently, this was the first time she had heard me speaking English, and it was like I was another person, one she did not know.

I had no suit. Katia and Vania drove me to a store that sold suits, especially those needed for marriages. They introduced me to a couple of clerks and left me so that they could do some serious shopping themselves. First, the clerks offered me a *batida*, a strong drink of fruit juice mixed with a native alcoholic drink made from sugar cane. Who was I to be discourteous? I accepted. Popular additions to the firewater included lemons, cashew fruit, and mango. The clerks offered me a peanut *batida*. It was so cold with small pieces of ice. It was delicious. The clerks brought so many suits for me to see. I hated trying on clothes, but as long as they kept serving me *batidas*, I would look at them. An hour later, Katia and Vania returned. They were surprised that I was having trouble standing, but I was very happy. They asked if I had found a suit. I had. They asked to see it. They were not happy. Apparently, Katia had not envisioned me in a blue, black, and white checkered suit. I had to start all over, and this time Katia and her sister stayed to stand guard.

THE YACHT TRIP

Natal had a yacht club located where the ocean met the river. The club itself was located on the river. Once yachts found the river, all they had to do was go upstream a little and dock. It was a small club because not many yachts were around, but one day a huge two-masted yacht came up the river and docked. My memory of its dimensions is foggy, but it could have been twelve feet by sixty feet. Its owner, a thirty-year-old Canadian who had his hair tied neatly in a long ponytail, never smiled.

We PCVs eventually met everyone who sailed into port. First, we were always traveling down the beach toward the yacht club, and second, the yacht captains and their crews were always seeking a nice beach located close to their yacht: that was our beach. It was inevitable that we would meet.

It did not take long for the yacht's captain and us to meet and end up in a bar trading stories. The Canadian captain had already spent a couple of years traveling around the world from port to port. He had just arrived from South Africa. He was going to Fortaleza and then on up Brazil's coast and eventually into the Caribbean Sea. After that, he had no idea.

He mentioned that he was one man short on his journey from Natal to Fortaleza and asked if any of us were interested. I was, but I had to get permission from my supervising agency. My supervisors happened to be okay with it since they did not want to be bothered with me anyway. I was going to Fortaleza by yacht

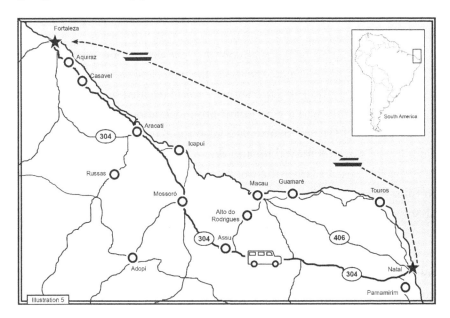

We sailed out of port late in the afternoon, with dark clouds covering the skyline over the water. We were sailing into rain. The captain took the first watch at the helm. It was his job to take us out to sea and set the course toward our destination. Another crew member was at the stove making supper. I was hungry. Yet another crew member was checking things on deck.

I was told that everyone had to take a turn at the helm and at cooking. The shifts were four hours long. There were eight bunks below where we could retire to wait for our shift, but I did not spend much time there initially. The motion of the yacht made me queasy and uneasy. I spent most of my time on deck, where I had the benefit of the sea breeze.

Sometime during the night, I was called for my turn at the helm. My orientation consisted of the captain pointing to a six-inch-long thick thread tied ten feet up onto a rope. The sea breeze kept it horizontal on the left side of the rope. The captain told me to keep it flying on the left. He added emphatically, "Never allow it to drop," and went below deck.

I did not feel very confident, but I did what he told me for four hours. There was nearly a full moon, and it was enjoyable to watch it and its reflection on the water. I loved hearing the yacht pound against the waves as it plowed its way north to Fortaleza. I was in heaven, even if I could not be certain that I wasn't may be about to send us all into a pile of rocks on the coast.

When it came to my turn at cooking, the orientation was equally as brief. The captain showed me the stove, the pans, and the store of food and condiments. Not being an experienced cook, I decided on scrambled eggs and rice. The stove had a wire guard around it about one inch from the stove's surface. I learned that was to keep the pans from sliding onto the floor as the stovetop moved up and down and sideways with the motion of the yacht. I used a larger-than-normal pan for the rice, so I did not have to fill it beyond half full, which allowed the water inside the pan to move about with the motion of the yacht without overflowing.

As for the eggs, I reasoned that if I heated the pan fully before pouring in the beaten eggs, they would stick to the pan before they had a chance to roll out of the pan and onto the floor. I was worried that the crew would complain because I was not a very good cook, but they did not. They ate quickly and quietly and even complimented me. They were just happy they had not had to cook.

After a couple of days, we sailed into port at Fortaleza. I shook hands with everyone, caught a taxi to the bus depot, and waited for a bus to leave for the eight-hour trip to Natal. I stayed seated while waiting because whenever I stood, the ground moved in a wavy motion that made it difficult for me to stand without putting my hands out to stabilize myself against invisible walls. People stared at me, so I sat down.

The trip back to Natal was not nearly as enjoyable as the trip up. It was very hot and slow. I just wanted to be back at my little house, where I could go into my backyard and use the *cuia* to take a wonderful bath from the water tank and then sleep.

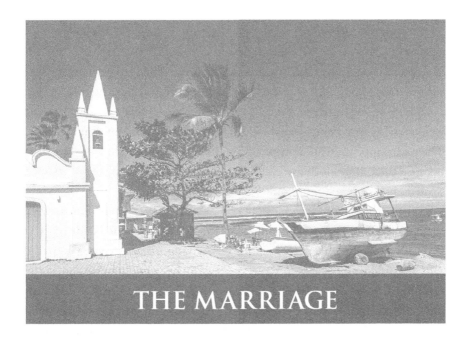

THE MARRIAGE

The day arrived. I awoke midmorning and showered. I put on my suit, but instead of dressing in my good shoes, I put on a pair of old, dirty tennis shoes. My best man came by to pick me up and drove me to Katia's house, where I intended to play a practical joke on her. Her house was long and narrow. On one side was the living room, kitchen, and an area to wash clothes and dry them on a line. On the other side, without the benefit of any corridor, were the bedrooms. The bedrooms were all connected, which reduced privacy, but it was the way the houses were constructed.

The house was bursting with people and activity. Apparently, the bride needed a huge support system. All I had to do was to exit my best man's car, and I immediately heard the gasp when one girl noticed my shoes. Her gasp spread to the next girl and the next until the sound went deep into the house. Then the gasps came back through the bedrooms until they reached the bedroom next to the street, whereupon I heard Katia shout, "Lyyyyynnnnn, take those shoes off this minute! Where are your dress shoes?" I responded that they did not fit, and all I had were my tennis shoes. Katia was starting to have a fit when I heard Dona Naide say, "Now what is important? Is it the shoes he wears or the act of getting married? He can get married barefoot. No one cares." I loved Dona Naide. She was my favorite person in the world.

My job was done. I returned to the car and exchanged my tennis shoes for my dress shoes, and we went to the church to wait for the bride.

Once we were at the church, I took my place near the altar. People were arriving and being seated. Soon, Dona Naide was standing near me. The padre appeared and reminded me that the bride had to arrive on time because there was another wedding after ours. We could have no delays. I smiled and shrugged my shoulders.

As the time approached, my best man and I were in position, as were the bridesmaids. The padre again looked at his watch. Dona Naide noticed this and told him that Katia would be arriving immediately. The padre said he could wait only twenty minutes before he would have to cancel the wedding. I told the padre that he was more than generous; I could wait only ten minutes. Then I would cross the street and have a cold beer at a nice outdoor bar located under huge shade trees. The padre shrugged his shoulders and announced that he would join me. Dona Naide pleaded with us to be patient. I felt guilty for pushing Dona Naide's buttons. She did not deserve that, but it was so easy.

Katia was ten minutes late. The wedding proceeded without a hitch until I was told I could kiss the bride. The photographer was there, ready to capture that important moment. We leaned in for the kiss, but at the last moment I pulled back and extended my right hand. When Katia saw my hand, she instinctively reached out to grab it, and we shook hands. Katia was a bit confused. The photographer took his photograph and spent his flashbulb. We then kissed, and the photographer swore loudly because he had not been able to replace his flashbulb fast enough to take the picture. He had no photograph of the kiss. But that moment of levity served to break all the tension that had surrounded the wedding. People started talking and laughing, and the fun began.

BACK TO THE USA—FIRST TIME

We were married on Saturday, and on Monday Katia sent her application for a new passport with her new name. It would take a couple of weeks to complete the process. Meanwhile, I treaded water while Katia made more and more purchases of shoes, dresses, and other outfits. She was nervous about going to a strange land without speaking English. I knew she was worried, although she said nothing. We were leaving her family behind for an indeterminate time, going to a place that she could not possibly imagine, to live a life that she could relate to only through the American movies that she had seen.

The day arrived, and we said goodbye to her family. She was filled with emotion. I was excited about starting a new life in the USA, but for Katia and her family, this was a time filled with doubt and the knowledge that they would not see each other for a long time. Her parents had to place their faith in me and trust that I would look out for and take care of their precious daughter. Dona Naide and Seu José's trust in me was enormous. I was grateful and promised myself that I would not fail them.

We entered the plane and took our seats; the door closed and was locked. Katia grabbed and squeezed my hand tightly, very tightly. Then suddenly, the pilot opened the throttle, and we were thrown back into our seats. On the flight from Rio to Miami, all messages were given in English and Portuguese, and half the passengers' conversations were in Portuguese. Katia had not yet felt the full impact of the English world that was coming.

After we deplaned in Miami, we were taken into a small room, since Katia was not a citizen. Within forty-five minutes, she had her papers legalizing her stay in the USA and her green card. Today, that process would take longer.

On the flights from Miami to Grand Island, no crew or passengers spoke any Portuguese, and Katia began to realize how difficult the language difference was going to be. In a few hours we landed in Grand Island. Mom and Dad met us. I could see that they were excited about meeting Katia, and I knew Katia was nervous about meeting them. This was when she fully understood that she was not in Brazil anymore.

When we arrived at my parents' home, there was a cold meal ready with all the trimmings. Both Katia and I were too excited and nervous to eat. We settled into our room and put our clothes away. We would be there for a few days before we drove to Columbia, Missouri, to buy a house.

Once we were in Columbia, we used the wife of a professor as our real estate broker. Her husband was the one who had helped me obtain my assistantship. Almost immediately, we found a nice house within easy walking distance of the school. On the first floor, the house had two bedrooms and a room for my study. In the basement, it had a full one-bedroom apartment with a functioning kitchen. We immediately rented out the apartment, and the income from that almost covered our house payment.

We spent several days having the utilities turned on and buying furniture. Katia loved the house. Buying furniture was not my favorite task, but Katia enjoyed it. After a few days we had selected the furniture, and it was delivered to the house. We were no longer sleeping on the floor and eating in restaurants.

Soon, the summer was over, and fall courses started. I had to work very hard. I was enrolled in a PhD program in agricultural economics, and candidates in the program were expected to have had a full course in calculus and to have a master's-level understanding of micro and macroeconomics and a working knowledge of statistics, including linear regression. I had none of that. My first year and a half were going to be spent taking courses that I should have already taken before starting my program. I had been admitted to the program due to my experiences in Central America, but I still had to compete with other PhD candidates who held master's degrees in economics or agricultural economics, and I had to meet the expectations of our Department of Agricultural Economics. My experiences in El Salvador, from my first Peace Corps experience, had gotten me into the program, but they would not keep me there. I had to meet the department's expectations at every stage in the program.

This department was of high quality. On the staff were a couple of professors who had been consultants to our presidents. One was the president of the national organization for agricultural economics professionals, and many had written books used widely across the country. Other professors were quoted frequently in the *Wall Street Journal*. This department had high standards and was highly respected by professionals across the United States.

Each course moved quickly, and I needed to study constantly. My life from 7:00 a.m. to midnight was either taking a course or studying for one. At night, Katia became sleepy by 8:00 p.m. and wanted to go to sleep. She had never slept alone, so she slept on the floor by my desk. I always went and got a pillow and blanket for her and made her as comfortable as I could. At midnight, I carried her to bed and covered her. I was awake soon after 6:00 a.m. and at the university by 7:00 a.m., studying again. On Saturday, I worked from 7:00 a.m. to 2:00 p.m. before going home to Katia.

Katia's life was simple but difficult. She had no way to learn English because I was never with her, and she could not drive without a driver's license. She could not obtain that license until she knew enough English to take the test. I was unhappy about having to go with her to the supermarket, but she could not drive or read the labels, plus it took a long time for her to make her selections. I was very purposeful and thought that we should be able to complete the marketing quicker. Because my life as a graduate student was demanding and involved inflexible schedules, I became wound tighter than a drum. This was difficult for Katia to understand and made living with me difficult.

It was not healthy that Katia spent so much time alone, but if I was to be successful in my studies, I needed to concentrate and make that sacrifice. I loved the learning. I loved calculus and statistics. I tolerated the economics. The professors moved quickly, and students had to keep up. No one slowed down if we were having difficulties. We were expected to visit the instructor for help or otherwise study enough to learn the material.

The course that I loved the most was linear regression. That would remain my favorite area of study for the rest of my life. It was so magical how we could take data and break it down into a series of conclusions. For me, the best present anyone could give me was a new set a data to analyze.

I learned that if a student in a master's program had difficulties, the professors were understanding and tried to help the student overcome those difficulties. If a PhD candidate had difficulties, he or she had to find a way to overcome them. The department threw no life preservers to the doctoral candidates. It was sink or swim. PhD candidates were expected to be able to solve problems. That would be the essence of their future jobs: to solve problems.

After three and a half years, I had completed all the required coursework and had to take the comprehensive exams. There would be three written exams and one oral exam: the first written test covered macro and microeconomics; the second covered the area of my specialty production economics; the third covered general topics; and then in the last exam, the oral exam, a group of four professors asked questions. Each exam was four hours long. We were allowed to fail and repeat one exam; otherwise, we would be released from the program and wished good luck.

These were the exams that we had feared from the beginning. These were the exams that had driven us to study those crazy hours. In the end, we had to pass them to move to the next phase: the dissertation.

I was fortunate to pass all exams on the first try. More than one student failed an exam and was required to take the makeup exam. I had a friend who failed the exam a second time and was unceremoniously removed from the program. That made it clear to the other candidates that the professors were serious.

Once I had passed the exams, I was cleared to start my dissertation. My advisor told me to find a topic for which I could obtain a grant to finance my research. I spent all day every day looking for a topic, but I found none. I was very frustrated. I had no experience doing a research project or finding financing for a research project. I started wishing that I had done a master's program with a dissertation instead of accepting the

PhD program with higher expectations all around. I was floundering. I went to my advisor for help. He tried to help, but I needed more assistance than he could give.

In the end, Katia informed me that she was pregnant. Surprise! I had not known that we were going down that path. It was my fault. She was tired of being alone. She informed me that she wanted to give birth in Natal, Brazil, where her family could help. I learned that pregnant women beyond six months along were not allowed to travel on airlines at that time. Not only that, but they were not allowed to travel with infants younger than three months old. This meant that if she went Brazil to have the child, she would be gone at least six months. I could not accept that.

I was not willing to be away from Katia for that long. I missed Brazil, and I had no direction in my program, plus I was burned out after three and a half years of nonstop studying. I notified the department that I would no longer need its assistantship and left the program. I was sad that I would not have a PhD, but I was very excited about our forthcoming adventure. Above all, I loved a good adventure.

We sold our house, packed our things, rented a truck, and drove back to Clarks, Nebraska. We arrived at midnight on Saturday night and spent Sunday with the family. We discovered that the doctor had found something in Dad's chest and was sending him to Omaha for more tests on Monday. Mom took Dad to his medical appointment while we started unloading the truck and returned it.

To our dismay, Dad was kept at the hospital. The doctor had found a malignant tumor in his chest and had scheduled surgery as soon as possible. Meanwhile, Katia and I had already sent tons of stuff to Brazil and had bought our airline tickets. If we did not follow through with our travel plans, all of our stuff would end up in Brazil without our being able to claim it. If we were not there to claim it, it would be lost since I had been granted a special one-time tax-free importation of belongings, but only I could claim them.

Dad had his surgery, and we spent as much time as possible with him in the hospital, right up to the moment when Mom and Uncle Wayne took us to the airport. Saying goodbye at the hospital was very emotional. Katia had a huge baby bump. Dad reached out and rubbed it for good luck. Katia had to be experiencing mixed feelings because she had been away from her family for three and a half years. I had no idea how long we would be gone or what awaited me. I had no job waiting for me. We were jumping into the darkness with all our money bet on a favorable outcome. I felt guilty about leaving Dad behind, but I felt I had no other choice. If we could have known about this even ninety days earlier, we could have aborted our move.

BACK TO BRAZIL

It was a joyful moment when we deplaned in Natal. When the plane's door was cracked, the air changed immediately. The heat and humidity exploded into the cabin, forcing us to recognize that we were back in the tropics.

We were taken to Katia's parents' house. They had cleaned the front bedroom, the one with the most privacy, and placed a full-size wood-framed bed in it. I was astonished by the wood. It was mahogany. I think the wooden frame weighed more than I did. The sideboards were three-quarters-inch thick. The end boards were thicker. Some of the decorative boards were three inches thick. I had never seen such a solidly constructed bed, and it was all pure mahogany.

First, we needed our own transportation. I bought a VW van, which in Brazil was used by stores to deliver merchandise. It was a commercial vehicle never used by families for transportation. Katia and her parents thought I was crazy, but her parents remained silent. That was their custom in our affairs: to remain silent. They never criticized me, even though I often gave them reason to do so. Katia was not silent, but I was adamant.

Katia did not understand that several years before I'd had a wonderful VW van named Zanderfeldt. I adored Zanderfeldt. We had gone everywhere together. Although the 1979 van that I had just bought in

Brazil had none of the amenities that my 1969 VW van had in the USA, it was still a van, and I loved it.

After a few days our boxes from the US started arriving. I used my van to pick them up and transport them to Katia's house. We started stacking them alongside the bed. The bedroom was not that large to start with, and with the large bed and now the boxes, we had to walk sideways to enter or exit the bedroom. In some cases the boxes were stacked four feet high. I was afraid they might fall on me while I slept or that Katia might give them a little push; after all, she was still sensitive about driving a VW van.

Our second goal was to find employment for me. In Brazil, you did not prepare a résumé and start spreading it around. You would never find a job. No one was hired based on his or her competence. People were hired based on who introduced them. The person who introduced you was referred to as a *costa quente* (hot back). My *costa quente* was a friend of Katia's family and was the editor of the largest newspaper in Natal. He wrote the best editorials that I have ever read. He had a wicked sense of humor, and when he applied it to someone, he could make the person seem ridiculously silly and stupid. No one wanted to be the subject, even obtusely, of one of his editorials. He could make the entire city laugh at someone; even that person's best friends would laugh because it was really funny. Everyone in the city either loved him or feared him, but everyone treated him with dignity and respect. No one wanted to invite his wrath.

We had arrived in Natal in November, and because it was so close to Christmas, no one was making hiring decisions until January. We were stuck in limbo. But Wolden, our editor friend, was reaching out to his contacts.

We started looking for houses. All lots were fifty by one hundred feet. I had to psychologically prepare myself for such small lots. Each lot had a wall built around its outer border for protection. Such walls were six to eight feet high, often with broken glass on top. Security was an issue, and you had to take measures to protect your family. Additionally, windows required metal bars for protection.

After looking at dozens of houses, we located one that fit our needs, although we did not appreciate the builder's choice of color schemes on the interior. The house had two large bedrooms in addition to the master bedroom, which included a large closet and bathroom. The floors were all covered in large, beautifully decorated ceramic tiles. A wide hallway featured the same beautiful tiles as the bedrooms. The kitchen had tile from floor to ceiling. The house boasted a large living room with dark wooden flooring and an office in the back. The living room connected to the garage, which was also all tiled. There, also was a maid's room and a place for her to do laundry by hand. We bought the house. The baby was due in early January. We were in a hurry.

In early January, Wolden came through, and I obtained a job at CEPA, the state agricultural planning commission where I had worked previously. It was a great job. It was considered to be a crème de la crème job in agriculture. I was starting at the top. This was great news since Katia was due any day.

I presented myself at CEPA for processing. The head of human resources was a very congenial man. I was given a small room at the end of a hall on the first floor. There were no windows, but near the ceiling, there was a hole in the wall to the outside that allowed some light into the room. I met my supervisor, who I discovered was great friends with Katia's family, but she was very cold and direct with me. I had the impression that she did not appreciate having a new employee dropped into her lap when she had played no part in the hiring process and when she did not need the employee. That was a hint of what was to come.

I walked around the office and found that my best man had moved to Brasilia, and a couple of others had also. I met several of the same people with whom I had worked while I was in the Peace Corps. They were communicative, which good manners required. But my experience in Latin America taught me to see that they were not happy that I had arrived. Their reasons may have differed: dislike of someone not already known by the group, dislike of Americans, or some other reason.

I was to learn that in general, Brazilian sociologists did not like Americans. I would meet many of them, including two who worked at CEPA, and they had a common, strong dislike for Americans. This was because of what our government had done to support their dictators at the expense of the common people. I'd had nothing to do with that, but they took it out on me.

One of CEPA's sociologists would not speak to me. We could cross paths in the narrow hallways, and he would always look down and away to avoid any need for communication. At first I was offended, but then I enjoyed it. I made him pay for his hate. Each time we were about to cross paths, I would speak a firm and happy greeting to him, making it obvious to everyone that he was being discourteous by not responding. Still, he never responded. I worked three years at CEPA, and he never once said good morning to me.

As I was introduced to CEPA employees who did not know me, I became used to the direction that the banter would take. They would ask questions to try to uncover who my *costa quente* was. They wanted to compare my *costa quente* with their *costa quente* to determine who was more powerful. They wanted to know where I had received my education and what degrees I had so that they could compare their qualifications with mine, to determine who was better trained. And finally, they wanted to know who my Brazilian family was so that they could compare socioeconomic levels. Everything involved a comparison. In the end, they wanted to know if they could treat me with disdain, if necessary, or if they needed to treat me with respect. I never gave away my *costa quente*.

Early on the morning of January 6, 1980, Kevin was born. Katia had no more than arrived at the hospital when he arrived. One year and two weeks later, Nicholas would be born. He would have much more hair than Kevin had at birth but a much weaker composition. Two years later, Christianne would be born, completing our family.

Back at CEPA, I was left with nothing to do in my isolated room for months. I took books to read on economics and calculus, or I practiced programming my HP41C calculator, the best calculator that HP ever produced, and I drank coffee—lots of coffee. For coffee, all I had to do was pick up the phone and order a coffee. Within a minute a man with a tray would appear at my door with a tiny cup of ridiculously sweet coffee. I bought a sixteen-ounce coffee cup, took it to work, and gave it to the coffee men. I informed them that I wanted it full and without sugar. This, of course, blew their minds, but they obeyed. After drinking such a large cup of coffee, I felt like I was blasting off to the moon. And I drank several cups each day, out of boredom if nothing else.

One day I was taking a break from my busy schedule, and I walked by a room in which a young person was on his tiptoes, trying to see out a small window located at the top of the wall. I entered the room and asked who he was and what he was doing. He was a college student studying economics and was working at CEPA as an intern. I asked him what he did. "Nothing" was his answer.

CEPA was given many different projects related to the state's attempt to plan agriculture. When a project came in, someone would grab an intern and have him make a series of tables. Often the same tables were needed in subsequent projects, yet each time, the tables were constructed from scratch. I went to the head of human resources and asked who was in charge of the interns. No one was. I asked to be put in charge of them. He thought that was a great idea.

I surveyed my colleagues as to what kind of tables they most often needed, and I prepared a list. I then handed out the assignments so that the interns could begin making the tables. Once done, the tables were sent to the typists to be made professional in a format that would allow them to be expanded as new data became available. These sheets were then placed in appropriately marked folders in file drawers. We were starting to build a data bank. Periodically, as new data became available, we just added it to the existing tables. The other technicians saw the benefit of such a data bank and started making more and more requests. As these requests were added, I was able to increase the number of interns.

I also taught the better interns how to use my HP41C calculator. I created programs needed to deflate prices and values to allow for comparisons across time. This was very important because inflation was 7 percent a month when I arrived in Brazil and increased slowly. Any comparison of prices across time required the prices to be adjusted according to each period's purchasing power of money. If I kept the actual prices in a table and the deflation index in my calculator, the actual prices could be adjusted to real prices in seconds and a new table created.

The students were happy to be learning useful skills that they could take with them to the real world after they graduated. More and more students were contacting CEPA's HR department to request an internship.

Northeast Brazil is noted for its periods of drought. We were still in the middle of a multiyear drought that was causing severe damage to the Northeast (an area at least as large as Texas) and its population. The World Bank came and wanted to view the damage. Since its delegation used English as a common language, I was sent to translate between the delegates and the people they visited. We traveled the entire state and were on the move for twelve or more hours a day. The heat was great, and dehydration always was a problem. I quickly started to suffer from severe migraine headaches because I could not eat and drink when I wanted, and I was constantly translating. Even during meals I was expected to translate, making it difficult to eat properly. I was miserable.

At the conclusion of the World Bank's trip, the delegation recommended that CEPA conduct a survey in the rural area that would allow the World Bank to quantify the damages caused by the drought. The World Bank would finance the study, but CEPA would have to create the questionnaire and collect the data. The sociologist who refused to speak with me was placed in charge of the study. At one point the team needed to determine how large a sample would be needed. The sociologist had to ask me for help. I enjoyed that task immensely.

My interns and I would process the data and create the tables summarizing the information retrieved from the surveys. While the committee was creating the survey, I was incrementally increasing the number of interns to seventeen. I trained them in the use of more sophisticated calculators. Once the preliminary survey was ready, we started preparing the tables that would be needed. These tables would be a guide to how we would have to summarize the data. Each intern received his or her assignment or was a part of a team that would go through the surveys and start filling in the blanks on the forms. After a few calculations were made, a more experienced intern checked for accuracy. I was proud of their work.

Within a week of the committee's completion of the surveys, the tables were finalized for the survey team's approval. I do not think I could do it any faster today with the help of a computer, although I could do it with less manpower.

TRIP HOME TO SEE DAD

I had been living in Brazil for six months, submerged in Brazilian culture. I never spoke English or heard English spoken. I watched TV in Portuguese, dreamed in Portuguese, and thought in Portuguese; in fact, I lived in Portuguese.

One night in early March 1980, I was awakened from my sleep by the phone ringing. I fumbled to find the receiver. I answered to hear my sister on the line, speaking English. I was taken aback because I had been in a deep sleep and because of the English. My brain was trying to process what was happening, and it could not. My sister was nervous because she had dialed a foreign country and was not certain that she had dialed the number correctly. She was frantic as she tried to drag a response from me, yet my brain was going in circles. Finally, I was able to say hello.

She told me that Dad was much worse and that I needed to come home to see him, fast. Mostly, I answered with yes or no. Maybe I uttered a short sentence or two, but I doubt the sentence structure was appropriate for English. Anyway, she seemed to understand that I had understood, and satisfied, she hung up. I knew what I needed to do. Further sleep was out of the question, even if it was the middle of the night. I started packing my bag.

When the sun came up, I was ready to go. I called Seu José and informed him of my trip. I had to inform CEPA that I would be gone for a while. The head of human resources understood and granted me permission, even though technically, it was against the rules. That was the beauty of Brazil—there were rules, but there was also the technique called *jeito*: the art of bending the rules. The HR director generously applied *jeito* to my situation and wished me luck.

I went to the bank and withdrew money. Seu José guaranteed me that Katia and Kevin would have no financial or other needs that Seu José could not handle while I was gone. He also drove me to see one of his friends who owned a travel agency. I bought a 2:00 p.m. ticket for a flight that would take me from Natal to Fortaleza and then on Belem and Manaus

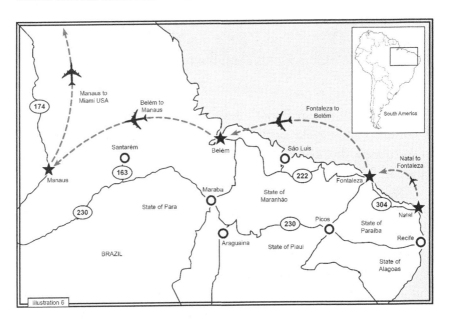

At midnight I would board a plane to Miami and then continue on to Atlanta and Lincoln.

Seu José and Katia saw me to the airport. I was frantic as I started my journey, wanting to magically appear at Dad's hospital. I checked my watch constantly. I hated each stopover and could not wait until we were in the air again. When I arrived in Manaus, I had a six-hour wait for my next flight to Miami. The hours were endless. I walked up and down the airport corridors. I sweated profusely because even with air conditioning, it was still Manaus, Amazonas. My body was tired, but I could not sit down, and I could not rest.

After an eternity, my flight was called. I grabbed my carry-on and walked outside toward the steps that led into the airplane. At the entrance to the plane, documents were being checked. I handed the man my passport, and he asked me for my exit visa. I asked him what an exit visa was. He told me that before I could leave the country, I needed an exit visa. I begged. He was emphatic. I returned to the airport to wait until morning, when I could find out what was going on.

I was lost and desperate. At night, time refused to pass. I saw only a couple of other people in the entire airport. They seemed as lost as I was. At 7:45 a.m., I grabbed a taxi and took it into the city of Manaus to visit the federal police, in an attempt to obtain my exit visa there. Again, I pleaded. The officials there were not impressed. They informed me that I had to return to Natal to obtain the exit visa there. They argued that my travel agent should have ensured that I had it before issuing my ticket. The travel agent had thought I already knew.

I bought tickets to return that afternoon to Natal. I called Katia and asked that the family pick me up in the evening. Once I arrived in Natal, I called home to Nebraska and informed my family of what had happened. They assured me that someone would meet my plane the next day. I was so wound up that I could not eat or sleep.

The next day, I learned that before I could obtain an exit visa, I needed to prove that I had paid all my taxes and that I had not committed any crimes. Seu José knew whom we had to talk to in order to obtain these documents. They were his comrades. I learned that normally it would take a week to obtain the required documents, but with Seu José and his friends, we obtained the necessary documents in hours. I bought tickets for the same route out of Brazil that I had used before.

After I received the exit visa, the trip proceeded without any problems. Upon landing in Lincoln, Nebraska, I grabbed my suitcase and found my ride home. I asked to go home first so that I could shower and change clothes, but Sally, my sister-in-law, thought it best that I go directly to the hospital. I knew then that Dad was extremely sick: that urgency frightened me.

I was led to Dad's bed, where he lay motionless. He was lying in the fetal position and seemed so small and frail. He must have lost a lot of weight. I barely recognized him. I did not know what to do. Mom told me to take his hand and talk with him. I did. At first it seemed uncomfortable, but then it became more natural. I described Kevin to him and brought greetings from Katia, whom he adored. After several minutes my exhaustion caught up with me, and my legs became rubbery. I asked if I could go home to sleep for a little and then return.

Sally took me home, where I showered and closed my eyes. A few hours later, I was awakened and told that Dad had passed. He had waited for me to arrive before letting go. I was disappointed that I had not been able to visit him again, but I was so thankful that I had managed to see him. I know that Dad waited for me to reach him and that it had not been easy for him to do so. I will be eternally grateful for those few minutes.

MILK STUDY

In the US, I never thought about milk. If I wanted some, I opened the refrigerator. If there was none, I went to the supermarket, where there were hundreds of gallons of all kinds of milk. I never doubted that a gallon jug actually contained a gallon of milk. I never doubted that if I bought 2 percent butterfat, it would contain 2 percent buttermilk. And I never dreamed that my milk might have been diluted with water. We could trust everyone in the supply chain.

I learned that in Natal, pasteurized milk, which was sold in one-liter plastic containers, was not trustworthy. You could not just buy it, pour it into a glass, and drink it like we did in the US. First, to me, it tasted horrible. Second, it had to be boiled because it might still contain dangerous bacteria despite having been pasteurized. Upon boiling, the milk often clumped. If so, it could be salvaged by adding sugar to convert the tainted milk into a form of candy. The milk we gave the children was powdered milk manufactured by a Dutch company. When hydrated, it tasted good, and it was unquestionably sanitary.

During the six-month dry season, the pastures dried up, and no grass was produced. The milk cows relied on hay, silage, or purchased grains for feed. The cost of producing milk increased, and therefore, the quantity produced declined. The producers faced difficulties in their attempts to create a profit because the government had established a price ceiling above which the price of milk could not rise. This price ceiling was often below the cost of production. The government forced a low milk price because many poor people could not otherwise afford to consume it. Even so, many families with children in Brazil still could not afford to drink milk, and a side effect of an arbitrary low price was that it discouraged producers from increasing their production.

There was one dry season when it was impossible to find milk in any grocery store or *bodega* in Natal. Both Kevin and Nicholas were drinking powdered milk at the time. Anybody who could find any milk would buy all the existing cans, leaving everyone else without any. In fact, I learned that people were paying grocery store workers to set aside cans of powdered milk for them when a shipment arrived. There was never any left to stock the stores. Powdered milk had disappeared from the supermarkets' shelves.

We were frantic. One Saturday I spent the entire day driving to small rural villages. My reasoning was that these villages were poor, and the residents seldom had enough money to buy milk, so some *bodegas* might have a can or two of powdered milk sitting on a dusty shelf. My trip yielded several small cans of milk. We never told other people how I had obtained the milk because I might have to do it again.

There were many sources of milk in the Natal region. First, milk powder was imported from another milk basin that produced a surplus of milk during the wet season. The best brand was produced by a Dutch company. This source was pure and sanitary. It had the exact weight and butterfat content advertised.

The second source of milk was another milk basin that produced a surplus during the wet season. That milk was sold by Parmalat in one-liter boxed containers with the exact volume and the exact butterfat content advertised, plus it was sterilized and perfectly sanitary. This milk had a flavor that was most like American milk, but it was expensive.

The rest of the milk came from the Natal milk basin producers. This milk basin never produced a surplus; even during the wet season, milk had to be imported. This milk was processed in many different ways. The first way was processing by the local milk cooperative. The cooperative pasteurized and marketed milk via supermarkets and a host of small *bodegas*, selling the milk in plastic bags marked to contain 1,000 milliliters with 3.2 percent butterfat. They always contained less than 1,000 milliliters of milk.

Most milk producers held back a portion of their milk to be sold on street corners in Natal as unpasteurized milk. Representatives of each producer chose the best corner they could and arrived with their milk in bulk containers in the back of their pickups. Consumers had to bring their own liter containers. They usually received a liter of liquid, but not of 3.2 percent butterfat milk—water had been added to the milk. The price for this milk did not follow the country's price ceiling; it was much higher, even though the milk had a lower butterfat content. This made the milk even more expensive for the poor people buying the milk.

I decided to conduct a milk study. Once a month, I visited the two largest chain supermarkets to learn how much powdered milk they had sold. I also learned how much the cooperative had sold. I then visited several pickups selling unpasteurized milk and bought their milk as any consumer would. I took each liter home and checked it for volume. I then took it to the cooperative, where the laboratory technician had agreed to check each sample for water content and for butterfat content. I recorded the price that each supposed liter of milk had cost me. I performed calculations, adjusting the price of milk from the various sources to reflect what I would have paid for exactly one liter of 3.2 percent butterfat, with no water diluting it. In the end, I had a weighted average price for the price of quality milk in Natal.

I conducted this study for more than a year. Each month, the real price for milk was increasing. The producers were making adjustments in the milk to increase the real price, to reflect the increasing cost of producing milk, in spite of the government-controlled price. The co-op decreased the size of a "liter." I found liter packages with no more than 670 milliliters—33 percent less than advertised. Imagine buying a gallon of milk and receiving only 2.7 quarts instead of four quarts. Imagine a twelve-ounce soda can containing only eight ounces. I never saw a single packet with more than 970 milliliters, and I examined samples of 250 each month for more than fifteen months.

As noted previously, some sellers of unpasteurized milk added from 10 percent to 60 percent water to the milk to enhance their revenue. Sometimes, the milk was so diluted that it looked pasty-white when poured into a clear glass. I was told that a few producers tried to cover this up by adding a little lime to the liquid.

Others skimmed the butterfat. Many of the samples had less than 0.5 percent butterfat. Most cows produced 5 to 7 percent butterfat in their milk. Thus, producers retained considerable value from the milk by selling it with only 0.5 percent butterfat.

I summarized my findings and presented them to CEPA, but the agency was not interested. The largest milk producers were the ones selling the most unpasteurized milk, and the owners of those companies were also the state's senators and representatives. Obviously, they were not interested in cracking down on the host of misrepresentations occurring in the milk market when they would be the targets of that investigation.

BECOMING SICK

It was Christmastime, 1980. Every year CEPA had a party at Christmas. We were all expected to attend. Even though I was never sociable, Katia and I had to go.

I had been tired for several days, but that night I was even more tired than usual. I was exhausted. Midway through the party, I had to beg forgiveness from our host. I was not feeling well. I had to go home to rest. I felt myself wanting to lie on the floor. I felt like my body weighed a ton. I nearly turned the driving over to Katia, but she still had a grudge against the van.

When we arrived home, I pulled off my clothes and lay facedown on the bed with my arms and legs stretched out. I felt dead. The next morning, I did not want to get up. I did not feel rested, and I had a severe headache. When I walked, I had to take tiny steps to minimize body movement and migraine pain. I could no longer lie down because it caused a throbbing pain in my head. I could rest only by sitting up in the bed. I was miserable. That was Saturday morning.

On Monday I called in sick to work and managed to get a doctor's appointment. Seu José drove, and Dona Naide balanced me on her shoulder. The doctor found that I had hepatitis. He ordered thirty days of absolute rest and said my diet needed to consist of only sweets. I could eat all the candy I wanted and piles of pancakes soaked in syrup, plus ice cream. I thought that the doctor must be crazy, but he was the doctor, right?

What a wonderful disease to have, at least once the headache was controlled. The only problem was that I was at the peak of contagiousness, and in addition to our having eleven-month-old Kevin, Katia was eight months pregnant with Nick. Happily, Katia's family jumped in to take care of me and keep me separate from Katia and Kevin. The doctor believed that I had contracted hepatitis from conducting my milk study. I had to suspend it.

PART-TIME CONSULTING

David, the American farmer in Rio Grande do Norte, had thrived while I was in the US working on my doctorate. He had teamed up with the Algodoeira São Miguel to produce cottonseed for them. The Algodoeira was owned by an English firm that was the world's largest producer of sewing thread. The company was fond of long-fiber cotton, which was produced in Egypt and Northeast Brazil. Most of the Brazilian production of this cotton occurred several states north of Rio Grande do Norte, but Rio Grande do Norte had the perfect soil and climate to produce the seed required for large-scale long-fiber cotton production.

David had access to large areas of irrigated soil located next to the Açu River. He also had the expertise to manage the large quantity of resources needed to produce a good crop every time. The Algodoeira and David made an agreement: The Algodoeira would finance the production and provide the land, and David would manage the resources and produce the crop. They had already been doing this for a year or so and were in the process of expanding the production area. David needed to know his costs and control his costs. For cash-flow purposes, he needed to know how much he had spent week by week. The Algodoeira was holding him accountable for the money they were advancing to finance the production, and he had to show the company where every dollar went and when it was spent.

I agreed to collect the data David needed. I created all the data forms necessary for such reports, and David allowed me to interview several employees to find some who were sufficiently literate and willing for me to train them to collect data.

I taught David's new record-keepers to fill in the forms each day. We needed to know what tractors and implements were used and how long they were used. We needed the same information on all labor. We needed to know about fertilizer, insecticides, and other chemicals. Cotton production was labor-intensive, so this was no small task. The employees were not used to writing anything down; most were illiterate and had no idea how important each detail was. Obtaining reliable numbers was going to be a large task.

Each Friday night, I traveled the three hours to David's farm, where I checked the records and worked with the employees who were collecting the data

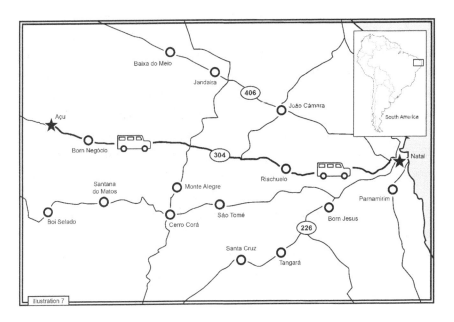

Illustration 7

It took several weeks before we were collecting data that was accurate. After a couple of weeks, I started taking preliminary reports to David. I checked the quality of the new data collected and returned on Sundays to Natal. On Mondays, I was back at CEPA.

This routine was very exhausting but necessary. Even my job at CEPA did not generate enough income for us to save any money. Our growing family required more cash. Brazil's inflation did not help. Inflation that had started out at 7 percent per month turned into 8 percent, then 9 and then 10 percent. Even 7 percent inflation monthly turned into 225 percent annually. Goods costing 100 dollars at the beginning of the year would cost 225 dollars by the end of the year. Our salaries increased only four times a year. During the last months of the year, we had to be on a tight budget.

David still did not have a house on the farm. He and I slept in hammocks strung up in a three-sided machine shed. The mosquitoes were horrible, but it was nice to sleep outside, even if it was hot. One Saturday night we got into our swimming trunks and went to the river to bathe. There was a gentle current. It was relaxing to stand with the water up to my lower chest. I allowed myself to float slowly downstream. It was heaven.

We put on clean clothes and went to David's usual bar in town, where he ordered beers. I was starving. After a couple hours had passed and David had drained several beers, I decided I would not wait for David any longer, and I ordered food. I had consumed only one beer but already had a severe migraine headache from lack of food and from spending the day in the hot sun. I should not have waited so long to order. I had been waiting for David's lead, but it had not ever come. I learned to fend for myself.

The most difficult part of cotton production was the harvest. It was all done by hand, and David had hundreds of acres ready for harvest. That year he contracted nearly 1,100 people to pick cotton. They came from all over the region, and many were not upstanding citizens. My problem was this: I had to formulate a method for tracking how much cotton each person picked each day. At noon on Saturdays, all picking stopped, and the 1,100 people would line up and expect immediate payment, in cash, and with exact change. Their goal was to hurry to the Açu market before all the good fruits and vegetables were sold. Before we could pay them, we needed to sum each employee's dumps of cotton, which we had only just obtained, into the wagon. It all took time, making the waiting employees restless.

To organize the labor, David selected twenty to thirty bosses, with fifty or so workers reporting to each boss. When the workers were hired, they were given a sequential number from 1 to 1,100. I printed these numbers on thick paper using my computers and then had each number plasticized.

A hole was punched through each ID and a chain strung through the hole to hang around the picker's neck, like dog tags in the army. Each time the workers dumped their cotton, they showed their number to their boss, witnessed the weighing, and initialed beside the record that they agreed with the assigned weight. We could not have discussions or disagreements about how much cotton had been picked on payday. Payday had to run smoothly, every time.

I needed to provide David with a list of how many bills and coins of each denomination he would need each Saturday. He needed this information by Wednesday. I had to find some way to estimate these numbers. For an added level of security, I needed to increase the number of smaller bills and coins to guarantee that no matter what a worker's final wage, we could immediately provide payment in exact change.

To obtain this quantity of cash, David would go to all the banks in Natal and take all the cash they could spare. His bank would then send a request to São Paulo to have the remainder sent on the next day's flight. This was a massive operation. Very few operations in Natal used 1,100 employees, and they could pay their employees by check. The cash payroll for 1,100 people was gigantic. It could also be a target for thieves.

On-site, David opened multiple tents to more quickly service the large number of pickers. Each tent handled a range of ID numbers. David had two men with machine guns guarding the money in each tent. It could not be underestimated how juicy a target for theft his operation was. A full week's payroll for 1,100 men could entice many people to try to steal the money. On Saturday, there might have been more cash on that farm than in all of Natal's banks.

Having the correct change under those conditions was difficult. After one Saturday, I convinced David to revise the next payday to include work done from Monday to Friday night. That would give me Friday night and Saturday morning to tally how much each worker was to receive on Saturday. It was not enough time to improve our early estimate for the quantity of bills and coins ordered from São Paulo, but we at least would know how much each person was to receive earlier.

The following week, the pay week was changed to run from Saturday to Thursday night. And then the next week, the pay week would be from Friday to Wednesday night. This gave us the chance to obtain exact change from the banks. In future years, the business would implement this procedure from the start of the picking season.

This project was a large one, with much money invested and flowing through David's hands. As soon as the harvest was done and the data collected, I presented David with his final report. I broke his operation into a dozen major categories and nearly one hundred subcategories. To avoid a massive report, I summarized costs into two-week periods from the beginning until the last cotton boll was picked. It was beautiful. I was very proud.

Eventually, David had to present this report to the Algodoeira. He was nervous, but he presented the same report I had given him, and they loved it. I was not aware at the time, but the Algodoeira had other producers like David in other states. All future years' contracts would require the producers to generate a report like mine for the Algodoeira. For me, that was the Good Housekeeping Seal of Approval.

DAVID VERSUS GOLIATH

David was a man of vision. He was a man who thought. I am sure that much of his best thinking occurred while he was lying in his hammock in his three-sided machine shed.

The state was in its second or third year of a drought. During years of drought, more black beans were consumed than were produced. People ate black beans once or twice a day, yet during a drought, production was far below normal. The state's inventory of black beans decreased during every year of a drought.

The government did not plan much, even if it had an organization called the State Commission for Agricultural Planning. We did only what the federal government told us to do. We did not spend much time thinking. No one cared enough to think.

Every new agricultural year, farmers needed black bean seed to plant. They always expected to have a good harvest, but in a drought they did not. The state's inventory of black beans decreased. One thing we could count on was that for each additional year of drought, the lower the state's inventory of black beans would be.

David knew that no one in the government was monitoring the inventory of black beans as it dropped more and more. David dusted off his technological package for producing black beans and planted all his irrigated land and all the irrigated land he could rent with black beans. He stored the beans and did this again and again until David was sitting on a gigantic inventory of black beans, far more than the state of Rio Grande do Norte had.

Finally, the rains returned, and everyone in Rio Grande do Norte and neighboring states wanted to plant black beans, but these states had far too few beans to sell to the farmers. David made it known that he had an abundant inventory of black beans. He offered to sell his beans at a price favorable to him, but the state of Rio Grande do Norte refused to pay. That was when a neighboring state came in and offered to pay his price and buy his entire inventory.

David would have made a huge profit in addition to saving the day for an entire state, except for one thing. The state of Rio Grande do Norte called an emergency session of its legislature and passed a law stating that in times of state emergencies, farmers could export agricultural products only with the state's permission. The government forced David to sell his beans to the state for a ridiculously low price. What a way to reward someone who just saved your population from a shortage of black beans!

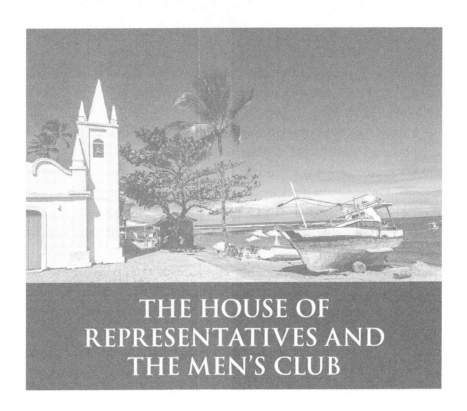

THE HOUSE OF REPRESENTATIVES AND THE MEN'S CLUB

In one of Wolden's editorials, he revealed information about the end of the year's legislative session. Apparently, the legislators rented two luxury buses, and after they ended their hard work, the entire legislative body went to the city's best whorehouse to relax. What's more, there was no public outcry of immorality. No one cared.

RUNNING ON THE STREET NEXT TO THE BEACH

When I first arrived in Natal after returning from the US, I tried to run several miles each day, driving to the beach to do my jogging. I was in good shape and wanted to stay in good shape. The only problem was that thieves learned that they could run up next to joggers and ask them for their wallet and their car keys. If a jogger didn't comply, the thief showed him or her either a knife or a gun. My wife was frantic every time I jogged. But of course, I felt sure they could not catch me.

If I saw anyone jogging toward me, I made sure to stay on or move to the opposite side of the street. If any jogger began closing in on me from behind, I switched to the other side of the street and ran faster. I tried to avoid running near buildings because a thief could pop out of a shadow and accost me.

Finally, it became too dangerous. I stopped jogging.

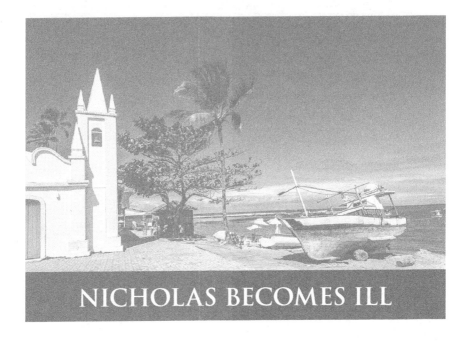

NICHOLAS BECOMES ILL

Nicholas was a small and fragile child, born at the beginning of the dry season. During his first months of life, it was very hot, and many babies were becoming sick and dehydrated. Soon, Nicky joined their ranks. We bought remedies at the pharmacy, and he seemed to get better, but then he would relapse. All the hospitals were filled to capacity. Newspaper articles announced that a baby had died while in the care of the state hospital, which had to accept anyone who appeared at its door. It was where the poor people went for medical care.

There were other stories of babies suffering from neglect at this hospital. This might have been because the staff was insufficient to care for all the babies who needed treatment. The administrators might have failed to increase staffing enough to deal with the increased demand. Or the staff might not have had much empathy for the babies. In Brazil, doctors were well known for having a God complex. They did what they wanted, when they wanted, if they wanted. The nurses and other staff might have copied this attitude.

One morning at 3:00 a.m., Katia woke me and told me that Nicholas was much worse. I looked at him; his breathing was very shallow, and he had a high fever. We called Dona Naide and told her. She woke Seu José, and they came in an instant to care for Kevin. Katia and I dressed and headed toward the private hospital where Nicholas's pediatrician worked.

We were turned away because the hospital was full. We went to the state hospital, where the staff told us we first had to drive across town to obtain a document from another office in the state's medical system. Begging did not help. We returned to our car and drove rapidly across town to obtain the required document and then returned to the state hospital.

When a male nurse came to meet me, he motioned for me to turn Nicholas over to him. I made it clear that I wanted to accompany Nicholas into the hospital and stay by his side. The male nurse said emphatically that this was not allowed. Hesitantly, I handed Nicholas to him. Before I let go, I told him that if Nicholas was not treated as he needed to be treated, I would find the nurse and punish him appropriately. This I promised him. I wanted him to know that he needed to take care of Nicholas, or he would be held personally liable. No excuses. It was very difficult to turn and walk away after seeing Nicholas disappear into the darkness of that cold hospital.

The next morning, we contacted our pediatrician. He found a spot in his private hospital, and we went directly to the state hospital and recovered Nicholas. He had been in its care for only six hours. It had seemed like a lifetime to me.

CHRISTMAS

Christmas in Brazil did not seem like Christmas. It occurred during the hottest part of the year, the time when people were into partying on the beach, and for me Christmas required snow and cold. It did not involve going to the beach. The stores had seasonal decorations, but that just made the celebration seem even more fake to me. Santa flying in a sleigh across the beach did not have the same impact as Santa sleighing over snow. I did not enjoy Christmas, ever, in Brazil.

BARBECUE ON THE CORNER

A main street running through Natal was only a mile from our house. On Saturdays and Sundays there was an outside chicken barbecue located on one of that street's busy intersections. It consisted of a man working under a huge tree, attending a large grill. He had a small refrigerated truck filled with hundreds of skinned chickens. Chickens were spread flat on the grill with their legs and wings going in opposite directions. The man was constantly basting and flipping the chickens. As they became done, he moved them to a higher level on the grill to keep them warm without burning them. The smell could be sensed two blocks away. It was impossible to pass the barbecue without stopping and buying a chicken or two along with a couple of extra-cold beers. Even now, I can taste and smell them. He sold hundreds each day.

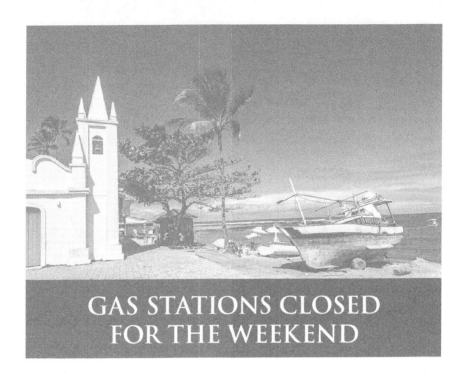

GAS STATIONS CLOSED FOR THE WEEKEND

When I arrived in Brazil, it was a time of economic growth. The government was trying to manage the economy, to make it do what the government wanted it to do. A very scarce resource was dollars, or foreign exchange. The government needed dollars to import machines for manufacturing, the driving force behind the economic expansion. The largest competitor for foreign exchange was the importation of petroleum. The government tried to force people to economize petroleum because everyone was driving around and buying gasoline, which squandered the country's precious foreign exchange.

To encourage its citizens to minimize their consumption of gasoline, the government set the price of gas at the pump at two or three times the price for which it was sold in the US. When the high prices did not encourage Brazilian citizens to reduce gasoline consumption enough, the government had to find another way. One morning we awoke to find that the government had passed a law forcing gas stations to close at 6:00 p.m. on Friday night and prohibiting them from reopening before 6:00 a.m. on Monday morning. This was designed to keep people home over the weekends, with their cars in their garages.

This new law immediately created long lines at the gas pumps on

Friday afternoons since drivers wanted to top their tanks off before the pumps closed. Monday mornings were also impossible since many people were running on empty, and long lines did not thin out until midmorning. The Brazilian people did not like this new law. *Jeito* would need to be applied.

This did not bother me since I did not drive much on the weekends, unless I had a trip to David's farm scheduled. When I did travel to his farm, it was his responsibility to fill my tank so that I could return home. All he had to do was set aside a couple of five-gallon cans of gas.

On my first trip to David's farm after this law was passed, Sunday afternoon arrived, and I mentioned to David that I needed gas to get home. He told me not to worry and climbed into my VW van's passenger seat. He guided me into Açu, and after several turns, we came upon a long line of vehicles. He directed me to enter the line. The line behind me quickly grew as we inched forward. I saw police cars both ahead of me and behind me. Eventually, we came to a wide gate in an eight-foot-high wall that surrounded a property encompassing an entire city block. I could already smell gasoline. I hoped that no one working there or anyone in the cars would be smoking.

Inside the walls I saw hundreds and hundreds of five-gallon cans filled with gasoline. There was an army of workers using funnels to fill cars, with as many as five or six vehicles being attended at any one time, including the police cars. Even the police were buying black-market gasoline.

This was another example of *jeito* in action. Anytime an obstacle was placed in a Brazilian's path, the first thing the Brazilian did was to look for a *jeito*, a way around the obstacle. A citizen could not undo laws passed, but he could find a place that sold gasoline illicitly. After a few months, the government desisted, and life returned to normal. The government had not taken into account the "*jeito* effect." Brazilians always found a way around an obstacle, like water found a way downhill.

TRIP TO PARAIBA FOR CONSULTING

David connected me to another small-scale American farmer, Mike, who had a small irrigated farm just inside the neighboring state of Paraiba, at the far end of Rio Grande do Norte

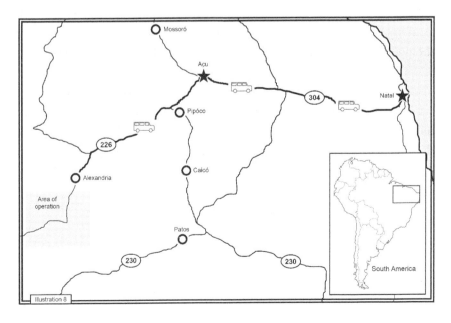

He produced milk, beef, pork, vegetables, rice, black beans, and fruit, all on a small plot of land. He wanted to know which enterprises were not making him a profit. He wanted to focus his scarce resources only on profit-making enterprises and remove the others.

I thought about what I needed to do. I created some forms that he would need to collect the data that I required in order to answer his question. When I was ready, I called him, and we agreed on a weekend for me to visit. Since it would be at least a six-hour drive, and since I had never traveled this route before, I left early Saturday morning rather than Friday night. This would allow me to drive exclusively during daylight hours.

Through most of Rio Grande do Norte, I had nice asphalt highways on which I made good time. It was hot and dry, and I had to stop frequently to drink water. After a while I spotted a narrow bridge ahead. On the left side, just a few feet from the road, were a couple of *bodegas* with several bicycles parked outside, like horses tied to a hitching post.

As I approached the bridge, I slowed and pulled into the center of the road. I looked in my mirror and, in the distance, saw another car following me. I slowed more as I approached the one-lane bridge. Then I saw something on the left side of my field of vision. I heard crashing sounds, and then an object was sent flying ahead of me on the edge of the road. I was entering the bridge, so I dared not look back. As soon as I had crossed the bridge, I looked back to see the car that had been way behind me stopped on the side of the road, just short of the bridge, with its front grill smashed and the bicycles strewn about. There were four big-boned men standing beside the car and shaking their fists at me, as well as men exiting the *bodegas*, shaking their fists at everyone as they tried to locate their bicycles. The car had to have been traveling at a very high speed.

Luckily for me, the pavement stopped at the bridge. The dirt road beyond the bridge was wide because the flat ditches were used as additional lanes. There were many potholes left from the previous rainy season. These had hardened into concrete-like surfaces, preserving the potholes until the next rainy season. In addition, large rocks rose up, by anywhere from a few inches to a foot or more, from the rock-hard clay surface. They were scattered about and embedded in the road. If a vehicle were to hit any of the rocks, the vehicle would be destroyed.

The men had reentered their car and had started their pursuit of my

VW van. To my advantage, the van had high road clearance, making it easier for me to go over the rough road. Their Chevette had low clearance, and given that there were four big-boned men sitting inside the car, the car had no chance of gathering speed on that road. I, on the other hand, put my accelerator to the floor and lowered my head to prevent it from banging into the ceiling as I sped on at top speed. They quickly disappeared from sight in my rearview mirror. They likely were still shaking their fists at me. Had the road been paved, they might have been able to catch me.

When I finally reached Mike's farm, he greeted me warmly. As I drank first a cup of cold water and then a cold beer, I told him about my experience on the road, and then we got down to business. I explained what I had in mind and showed him the forms that I had previously created. A few changes were made as we reviewed them.

I discussed an elaborate experiment that would be needed to properly determine whether producing milk could be profitable. It involved modifying the cows' diet and would take three months to execute. Mike agreed. I did not want to ask him to eliminate his dairy herd if I could not be certain that it was the correct decision.

I returned after a couple of weeks to check on the collection of data and to answer any questions that Mike had. I was very cautious as I reached the narrow bridge and the dirt road that followed. I kept my eyes glued to the rearview mirror for any cars traveling at high speed.

In the end, I advised Mike to eliminate the dairy cows, the beef herd, and the pork production. The grain, vegetable, and fruit production enterprises were all profitable. Mike seemed happy. I think he had already known that this would be the case. I returned home. My job with Mike was done.

CONSULTING FOR THE ALGODOEIRA

The Algodoeira São Miguel had been operating in Brazil for nearly 150 years. During this time the company's people had collected seeds from all the different native cotton varieties they could find. They took meticulous notes on the characteristics of each variety and each year crossed different varieties, attempting to find new varieties with improved characteristics.

As time passed, this work grew more and more complicated and expensive because they had to find ways to safely store the seed so as not to lose any of the varieties. Finally, in 1981, they had to let some of the varieties go. It was too expensive to maintain all of them. Normally, this job would have gone to the company's in-house statistician located in the Algodoeira's main office in England, but she had a backlog of projects that she could not complete for a year or more.

This was when the Algodoeira contacted me. I explained that I did not have a PhD in statistics and might not be able to complete the task. They wanted me to try. They handed over all the data to me. Luckily for me, during my doctoral training, I had taken many statistics courses. One was a course normally not studied in the field of economics, but I was interested in agricultural experimentation, so I had taken the course.

During the last week, the professor had introduced a methodology that had just been developed. In fact, he had just read about it in a publication. He had copied the table needed for us to apply the methodology and handed it out to all students. I had saved it.

Applying this methodology, I tested all the seeds and made my selections for seeds to be retained and seeds to be abandoned. I explained the methodology to the Algodoeira and gave them my findings. Later, I was informed that their statistician in England had unexpectedly found some time and had examined the data. She had applied a methodology similar to what I had used, and her selections were very close to mine. I did not mind having my work checked. I felt the pressure of making a mistake and causing the dumping of a 150-year-old seed that might result in it disappearing from this earth forever. I was happy to share the blame if we were wrong.

OPENING A COMPUTER STORE

My consulting clients needed a quick turnaround for reports. As soon as they turned over the last piece of data, they wanted their final report. This required me to have my computers and printers working all the time. In 1981, both computers and printers were prone to not work. The computer that I used had a stolen reverse-engineered version of Radio Shack's TRS-II, or something like that. It had two weak points: its hard drive (five-inch floppy disks) and its keys. The drive constantly needed alignment, and it was common to have two or three keys not working on the keyboard. I needed in-house computer repair.

I had already opened a business for my consulting: ECONSULT. If I rented a space and hired a repair technician, I would have my in-house technician, but he would be expensive as he waited for my computer to break down. I decided that I might as well sell computers and—why not?—offer programming classes. Computers were just starting to be used, and more and more people wanted to learn how to program.

I rented a large house near the city center. I bricked over the garage door opening, repaired the inside, and added an air conditioner plus a sturdy eighteen-inch bench around the walls to hold the small computers and TV monitors. Beside the computer programming classroom was an old room previously used to store tools. I remodeled it, added an air conditioner, and called it my office. The large living room was arranged to showcase computers, printers, calculators, and accessories. Another room was used to store spare parts and for the technician to make repairs. Still another room was used by the accountant and his staff. I allowed the accountant to use his room for his outside business in exchange for his free service for my company.

I advertised for a programming class to start on Monday morning and flew to São Paulo on Wednesday afternoon to acquire the right to sell the most popular brand of computer produced in Brazil. I bought one unit of the company's large computers and one unit of its medium computers. Then I bought twelve units of a tiny computer that would be used in the beginning computer programming class.

When I returned home on Friday night, all I knew how to do was turn on the computers. I had no software for any of the computers. The computer company had included, for free, the game of hangman. I parked the large and medium computers and grabbed one of the small computers to be used on Monday for the programming class. I had bought a book explaining how to use the BASIC programming language. By Sunday afternoon I was beginning to get the hang of it and started preparing my first lesson.

I was ready for the first class on Monday morning. I was relieved when that class went well. In fact, the rest of the course went smoothly too. New courses formed quickly, and I had to hire professional programmers to teach them. We gained a reputation for being honest and offering high-quality courses. It helped when Brazil's gigantic petroleum company started sending its employees to us for training.

The professional programmers organized a course in COBOL using our midsize computers. This was also a success. Companies had an increasing demand for data-entry workers. I created a course to train students in data entry. It was a self-taught course where people could come in anytime, as long as they had scheduled time on the computer, and follow the directions. The lesson plan was carefully mapped out; all the students had to do was follow it. When they felt they were done, they would take a timed test. Afterward, we provided them with a document showing they had trained with us and showing their typing speed and the percent of correct numbers entered during their timed test.

Over time, our courses were what maintained ECONSULT and

allowed us to pay our bills. Computer sales were nonexistent. I blamed myself because I was not a salesperson and did not know how to hire or supervise salespeople. I was a teacher. I should have focused on my strong suit instead of my weak points.

Our third source of income should have been equipment repair, but no revenue was forthcoming there either, in part because we had to sell the service and our sales department was ineffective. Eventually, I learned of another reason there were no repair sales: my technician was often contacted for repair, but he had the clients pay him directly rather than include me. For parts, he asked me for computer boards that he said were needed for the maintenance of our computers and then robbed parts from them to keep his side business going.

When I discovered what was going on, I fired him and found a young man who had just graduated from technical school. I hired him and, at great expense, sent him to São Paulo to become accredited in the repair of our computer brand products. I could trust this young man, but much damage had already been done. I had him go through our inventory of computer boards to determine how many had been robbed of parts. To repair these boards, I had to pay more than two thousand dollars.

TARANTULA MATING SEASON

At home, I usually spent each evening in my office. I studied computer programming, calculus, or other topics that had caught my fancy. One such night, Katia came to the doorway but said nothing. When I looked up and saw her, she was pointing toward the ceiling. I looked up and saw on the ceiling directly above my head a tarantula that had to be six to eight inches across.

I ran for the broom, knocked it to the floor, and attacked it. Katia grabbed the dustpan. I swept it into the dustpan and ran it outside, where I dumped it on a concrete slab. Katia handed me the cigarette lighter fluid and matches. She pointed at the tarantula blob and then at the lighter fluid and matches. That was the Brazilian way. I obliged her.

Katia informed me that January (it was January) was the month that tarantulas chose for mating. She told me to check our outside wall. She gave me the lighter fluid and matches to take with me. I put on sandals and started to check our perimeter. Before I was done, I had found six or eight tarantulas climbing down the inside portion of our outside wall into our property. Perhaps I should remind people that our house was the only house standing on three or four city blocks. It might have been for that reason that we had so many tarantulas invading our property. Even so, after a few days, mating season must have ended because the invasion of the tarantulas stopped.

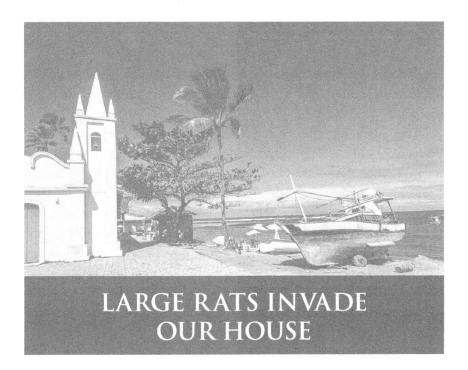

LARGE RATS INVADE OUR HOUSE

The area around our house was flat, and our house was the only house on four square blocks. There was much unoccupied ground nearby. The city had not even marked off the streets. There was no need until people started to occupy the area, and the city did not have the budget to provide services before they were needed.

Natal had many homeless people. They usually banded together in an area and created a shantytown with homes made from cardboard or scraps of wood, with bits of tin serving as roofs. The houses were tiny and fragile, but they offered some protection for their inhabitants. Normally, these people avoided property lots because the lots' owners were very protective and would sue the city to have the squatters removed. The streets around the blocks, however, were another thing. They were owned by the city, and the city never knew what was going on. These shantytowns tended to follow the streets. It was very unlikely that anyone working for the city would care enough to create a fuss and have a shantytown removed.

Such a shantytown appeared on a street one block from our house. It grew a little each day and became a significant hideaway for homeless people. One day the city placed signs throughout the cardboard homes, stating that the residents had to leave by a certain date. Some inhabitants left, but others did not believe the warning and stayed. On the stated day, bulldozers appeared, and within an hour the town no longer existed. It was gone.

That night Katia kissed the heads of Kevin and Nicholas, both of whom were still sleeping in cribs, and went to bed. Later, I came out from my office and also went to bed. During the night, something woke me. I sat upright in bed and listened. I heard nothing, but something was not right. I crept over to the bedroom door and turned on the light. Just above my head, walking along the ledge of a slatted area above the door, was a giant rat. I do not exaggerate—from nose to tip of tail, he was no less than two feet.

Katia turned and sleepily asked me why the light was on. I told her to look at the door. She screamed. The rat dropped to the floor and disappeared down the hallway. Katia yelled for me to close the door to the boys' room. I hurriedly closed their door and also the door to the third bedroom. I collected the large glass cover from a coffee table and placed it across our bedroom door opening, impeding the rat from reentering our bedroom, where Katia had taken refuge and from where she yelled encouragement. Then I went to the kitchen to grab a broom and made sure that the bathroom door was open. I was ready for battle.

I found the rat in my office and herded him from there, through the living room and into the hallway leading to the bedrooms. I beat the broom against the floor to rattle the rat. He picked up speed as he headed toward our room. He hit the invisible glass table cover that I had placed in the doorway. He lost his senses just long enough for me to catch him and force him into the small bathroom between the bedrooms. I jumped in behind him and closed the door. It was him and me, *mano a mano*, in a tiny bathroom. Katia yelled encouragement from a safe distance. The broom was too long for me to swing it properly, but I was able to use it effectively enough to doom the rat. I got the dustpan and threw him over the wall. I would deal with him tomorrow.

Katia and I thought the rat had come from the destroyed shantytown. Now the rats that had lived there also were homeless. The next night, I had to deal with two more rats, and each night thereafter for a week or so, I had to fight it out with more rats—and then it stopped. The rats had moved on and likely found new housing elsewhere. Our lives returned to normal.

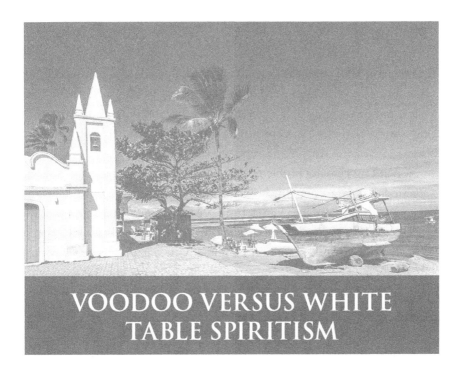

VOODOO VERSUS WHITE TABLE SPIRITISM

Anyone who lived in Brazil needed to be aware of voodoo because it was widely practiced. Before I moved to Brazil, the only thing I had known about voodoo was what I had seen in the movies—not much. In Brazil, voodoo was called *Macumba,* and the *Macumba* priest was called a *macumbeiro*. *Macumba* was used to communicate with spirits, usually to bring about bad outcomes to one or more people. The *macumbeiro* did not care. Whether he or she was asked to do good or evil, the *macumbeiro* just did it—for a price. The client always had to pay for the *macumbeiro*'s services.

There was another form of Spiritism as well—White Table. This branch of Spiritism was very different from *Macumba*. First, it was so named because the host sat at a small table covered in a white cloth with a Bible on one side. White Table priests never would do anything that could bring harm or distress to any person, and they never charged for their services. They believed that their gift of communication with the spirits was a God-given one; therefore, it should be shared for free with anyone who needed it.

In Brazil, Spiritism (White Table) was an official religion. In fact, even though 92 percent of Brazilians were Catholic, 44 percent were Spiritists.

You may ask how that was possible. I asked a person who claimed to be a member of both religions, and he said, "Clearly, I am a Catholic, but what if my time comes and I am at the pearly gates only to discover that God is not a Catholic but a Spiritist? That is why I am also a Spiritist—for insurance, because surely God is one or the other."

When I was having problems with ECONSULT, Vania, Katia's sister, suggested that we accompany her to a White Table session. I was reluctant but curious. I agreed to go. Vania, from previous experience, knew that we needed to take with us a bottle of *aguardiente* (firewater) and a vile, cheap cigar. We went to the practitioner's home with our purchases. A white-haired lady in her mid-sixties met us at the door. Her house was neat, with many religious symbols displayed on her furniture and walls. She led us into her office, where she had a card table set up. It was covered in a white tablecloth, with a Bible placed neatly at one corner.

Vania introduced Katia and me to the woman. She was very gracious. After learning that this was our first White Table experience, she explained what was about to happen. She explained that the spirit communicated through her, but when the session ended, she would have no recollection of what had been discussed.

We held hands while she prayed. After her prayer, she lowered her head and muttered words I could not hear. She was calling the spirit with whom she preferred talking. He was an old African slave from the early 1800s. When he appeared, he always asked for strong "firewater" alcohol made from sugar cane and cigars made from the strongest and most vile tobacco every produced. When he lived as a slave, these would have been the best he could have hoped for. The spirit's name was Preto Velho (meaning "old black man").

When his spirit came, our host's body jumped like someone had hit her really hard on her back or like she had received a large electrical shock. She continued to look down with her eyes closed. At this point, she had no idea what was happening. We knew that Preto Velho's spirit was in control when our host gave the belly laugh of a happy old man who was about to receive a gift. Occupying our host's body, the spirit immediately asked for his firewater and cigar. He took big swigs of firewater and smacked his lips and inhaled a large drag from the cigar. Within a couple minutes the room was filled with smoke. We opened all the windows. Perhaps I should note that our host never smoked or consumed alcohol—ever.

After Vania had discussed her business with Preto Velho, he paused, and then stated knowingly that I had questions that I wanted to ask him. He told me not to be afraid, to ask him anything I wanted. His answers were often funny but spot-on. He was a gregarious man, one I would have loved to meet in the physical world and call my friend.

Suddenly, he said someone else was calling him and he needed to go. Within a couple of seconds, our host's body snapped back and forth, and he was gone. The woman would need three or four minutes to collect her thoughts and become her normal self. Meanwhile, we had to empty and hide the ashtray and the firewater bottle. We had to try to disperse the smoke. She had no idea that she consumed alcohol and smoked a vile cigar each time Preto Velho came.

When she was able to reorient herself, she asked us what had transpired. She showed no sign of having consumed such a large amount of powerful alcohol, and she was not aware of the smoke residual that should have been in her mouth but was not. We waited and let Vania speak because we did not know what we could and could not say. After our host said another prayer, we thanked her for her generosity and disbanded.

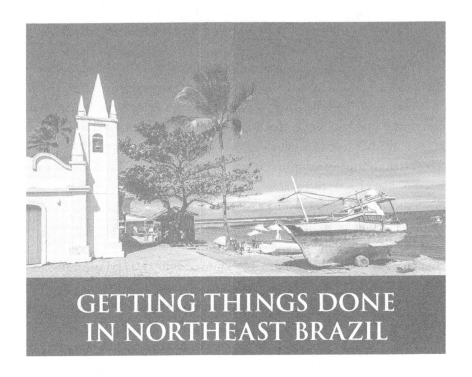

GETTING THINGS DONE IN NORTHEAST BRAZIL

Brazilians, or at least *nordestinos* (people from Northeast Brazil), never did business with a stranger. That was me, the stranger. They always did business with family or friends, even if the price offered by the stranger was more favorable. In northeastern Brazil, you could trust family and friends—and no one else. Friends were not made overnight. It took years, maybe decades, to establish friendships, and a few marriages somewhere along the line also could help.

Economists, especially marketing people, would not understand the *nordestino* economy, where people would pay more for something even when it was offered for less by another person. Most economists would find the *nordestino* an irrational being. But after living in this strange environment for years, I came to understand it.

Once I visited a business and found it receptive to buying a computer. The owners asked me to present them with a proposal and said we could talk. I returned to my office and prepared the proposal, which I left on top of my desk because I expected to return to the business to present my proposal yet that afternoon.

As soon as I printed the proposal, a young man entered my office and sat on the corner of my desk and started fidgeting with the proposal.

I had seen him around Natal. He made money doing favors for friends, but he had no job. His family was middle-class, even though he presented himself as being from the upper class. He looked like he had just left the tennis courts at the country club. I should have reached out and removed the proposal from his hand. It was evident that he was reading what was printed on it. He suddenly told me to pay him a commission of 10 percent and he would close the deal. I denied his request, telling him that he was not an employee and could not represent my company without being an employee. He said that he would speak with the firm's owner and "put dirt on the deal" so that I could never sell anything there. This irritated me, so I told him to do what he must. He did. I was never again able to get beyond the front desk at that business. I do not know what he told them, but he made good on his promise. I never again allowed him into my office.

Doing business in Natal was complicated in ways that an American could never imagine. For example, when I needed to obtain a driver's license, I was told to go to a government office between the hours of 10:00 a.m. and noon and wait in line, and this was only on Tuesdays and Thursdays. I arrived at 10:00 a.m. to find a long line. At noon, I was still a few people from the counter, but at exactly twelve noon, the counter employee slammed a "closed" sign onto the counter. Without any explanation, he turned and left, leaving many unhappy people in the line.

I learned to arrive before 8:00 a.m. so that I could reach the counter before noon. I was elated when I reached the front of the line before closing time. But the counter agent told me that I needed to fill out a form before I could apply for the driver's license. He explained, with great remorse, that they did not have the form at this office and that I could find it in a bookstore. I went to three bookstores before I was informed that there was only one bookstore in the city that carried the form. It was completely across town. When I found this store, its clerk informed me that the store was out of the form, but next week they should have more available. He showed great compassion before he turned to help another person.

I returned to the bookstore twice the next week, but they didn't have the necessary form either time. On my third trip, I found that the form had arrived, but unfortunately, I could not pay for a government form there. I had to pay for it at a specific bank branch located in another part of the city. With great difficulty, I found the location. I was surprised that all I had to do was stand in line, pay, and obtain my receipt, which I did without difficulty. I returned to the bookstore to discover that the receipt had to be stamped by someone at yet another office located in another corner of the city. That stamp took only one more day to obtain. With the proper receipt, I received the form with ease. On the next Tuesday, I was in line at 8:00 a.m. When I reached the counter, just before noon, I was informed that unfortunately, the director had had to leave unexpectedly, and only he could sign a document that I needed. The counter agent suggested, with obvious satisfaction, that I return at a more appropriate time. He reached for the "closed" sign, slammed it onto the counter, and disappeared through a door.

It was then that I mentioned my problems to Seu José. He offered to drive me to the office after lunch. I reminded him that they were open only between 10:00 a.m. and noon on Tuesdays and Thursdays. He smiled and told me not to worry. He might find a *jeito* to resolve my problem. *Jeito* was very important. You had to have it to get things done, yet it could not be purchased. It was more important than money and much harder to obtain.

Seu José confidently led me into the government agency that Tuesday afternoon. He smiled and spoke familiarly to everyone as he walked straight to the counter, although there was still a line for something else. He looked through an open door into the director's office. The director saw him and waved him in. We went around and behind the counter and through the director's door and shook hands all around. The director asked what Seu José needed. He told the director I needed a driver's license but had encountered a few problems along the way. The director was very sorry for all my difficulties and called to an employee to come and bring the form with him. The employee appeared in the doorway with a form. The director handed me the form and requested that I fill it out. He even gave me his personal pen to use. In five minutes I had my driver's license.

That was *jeito*. Seu José had it because he had carefully cultivated many friendships during his decades living in Natal. He had helped many people with problems and never asked for any payment for his services. Payment would come later, when he needed a favor, like now. That was how a *nordestino* survived. He made friends and kept them by doing favors when requested. Never should a favor be requested for minor things. That was wasteful. Seu José had just cashed in one of his favors for my driver's license. No one kept score verbally, but everyone knew the score. You never discussed it, but it was always understood when a new favor was owed and when an old one had been cashed in. Seu Jose did this for me because I was his son-in-law. I was family.

In spite of my personal lack of *jeito*, I made a modest living by helping large farmers calculate their production costs and by preparing budgets for them. Most of my revenue came from my series of computer classes. We were always finishing a few classes and starting others. They were always filled to near capacity.

No one else in the city yet offered computer classes. Life was good for a few months, and then one day, a class ended, and no one signed up for another class. Then another class finished, and again, no one signed up for another one. Finally, all classes were over, and I had no income and no cash reserves, since by this time, I had left the employment of CEPA, I was in trouble.

I had been living on the edge for years. The problem with living on the edge is that sometimes you fall off. I was worried and did not know what to do. The transition from surviving to not surviving was quick and painful. I had never contemplated that this could happen to me. I did not understand. I was living in a foreign country and had no one I could turn to for financial support. My father-in-law had a small amount of *jeito*, but he was a simple man with limited financial resources. He could not save me with money. It was a critical point in my life. I had to find a way out—and fast. I could not involve Katia's family in a problem that was not theirs.

In Nebraska, life had been easier. My banker there was the son of my father's banker and the grandson of my grandfather's banker. There was equity in this relationship. When the banker loaned me money, he knew that my father and my grandfather had never failed to repay a loan; therefore, it was easier for him to extend a loan to me. In Natal, I was a foreigner with no capital and no equity, and even my father-in-law's reputation did little to help me obtain loans. Furthermore, short-term loans cost from 20 to 24 percent a month while inflation was between 10 and 15 percent per month. Interest was not your friend.

One night, it was late, and the store was closed. I had been in my office, trying to find a solution, but I finally decided to go home. At the same time, my night watchman was arriving for his watch. We needed a night watchman because we could not leave our computers and electronic goods exposed to thieves. Instead of going to his room, he approached me. He was a simple man who came from a rural village. He was not comfortable speaking with people outside his social class, but I had always been good to him and always spoke kindly to him. His eyes were directed toward his feet when he spoke. He said, "Pardon me, Mr. Lynn, but I noticed that there are no more classes. Are you not offering any more computer classes?"

I explained to him what had occurred. He listened, but I felt that he had already known what was happening before he came to me. He looked at his feet again and paused before he suggested that this might be a "work." I had no idea what he meant. My Portuguese was excellent, so I understood the literal words he had said but not his meaning. I asked him to explain. He said it had to be a "work," a spell, a curse placed on me by someone who did not like me or was envious of me.

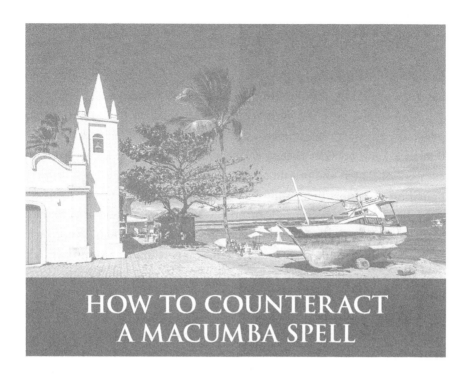

HOW TO COUNTERACT A MACUMBA SPELL

The night watchman told me that he knew someone in the country who could remove the spell. I told him that we knew someone from the White Table who could handle this. He begged forgiveness but adamantly claimed that White Table magic was not strong enough to overpower *Macumba* magic. He said, "Sir, you can't fight fair in an unfair fight and expect to win." He continued, "It takes a mighty powerful *macumbeira* to overpower *Macumba* magic." He told me that the priestess he knew was very powerful and could handle the job. I told him to make the appointment. I could not sustain another week without computer classes.

The next day the night watchman, Katia, and I drove two hours to a small village located off the highway. We turned off the highway onto a steeply downward-sloped cobblestone road. The road was filled with holes from the cobblestones being washed away during heavy rains. We headed downhill and into the village, advancing slowly to keep from smashing the bottom of our Chevette, a car I had recently purchased, on cobblestones each time a tire dropped into a hole. After a couple of blocks, the cobblestone stopped, and dry clay continued, also with deep potholes left from the last rains. At the end of the village, we turned left. There were no houses on our right. Mid-block, there was a large open entrance into an area enclosed by a six-foot-high wall of adobe bricks.

We stopped near the entrance and continued on foot. No one was inside. There was a large open area covered with a roof made of palm branches. On two sides there were benches for people to sit on, as in an auditorium. Later, the priestess arrived and assured us that people were preparing. A couple men walked in with small drums and set them up near one bench. Ladies started appearing. They were dressed as Bahianas: black women from the state of Bahia dressed as the women would have dressed 150 years ago. Each woman had a scarf wrapped around her hair and wore a white blouse and a skirt with many layers, similar to the skirts used in US square dancing.

The drums started beating a very infectious rhythm. One by one, the women fell into a line and started dancing in a large circle. They twisted and turned as they went about in a wide circle and held their arms up. Inside the circle were chalk markings on the concrete floor. Suddenly, one woman dropped to the ground and started flopping around like a fish on dry land. A couple of other women stayed with her to keep her from hurting herself, and then another woman dropped to the ground, and others stayed with her. The priestess went to each woman and spoke quietly with her until she could stand and dance again.

Our night watchman explained that dark spirits had come uninvited and occupied the dancers. The priestess sent these unwanted visitors on their way as quickly as possible, but they did not always want to leave. The women continued their dancing until they could attract the spirit they wanted for this task. After more than a half hour, the night watchman told Katia and I that he thought they had contacted the spirit they wanted.

At this point, someone appeared with two or three black chickens. The priestess cut their throats, and their blood was drained into a bowl. It was then sprinkled about, forming designs on the concrete. While the priestess was busy doing her stuff, the other women continued dancing in a circle and singing. Katia and I were shocked. We'd had no idea this was going to happen. We were very uncomfortable. Suddenly, the ceremony finished.

The priestess came to us. She said that her work had been successful. She told us that we needed to "sweep [our] roof, every day." Our night watchman assured us that he knew what to do. She told us that somewhere in our house, probably in the far-right corner, we would find five coins buried vertically. We needed to find them and carefully extract them without touching them. They had to be placed in a glass jar and covered in frog urine. The jar had to be covered and buried in a special manner. Our night watchman assured us that he knew what was needed. I should state that this simple priestess had never been to Natal and had no idea what business we were in or what our store looked like.

We paid her and headed home. Once we were back in Natal, I drove to ECONSULT. We started looking in the kitchen, which was located in the far-right part of the house. Under the sink there was a concrete floor, but the concrete was of two shades. It appeared that a part of the concrete had been broken out and then replaced, although this had likely occurred some time ago since the concrete was not fresh. I grabbed a hammer and started pounding. I found five coins, each placed in a vertical position. The night watchman took over and disposed of these coins as asked by the priestess. I have no idea how he found the frog urine. I prefer not to think about that.

The next morning, very early, I climbed onto the roof and found no fewer than six small packages wrapped in plastic. Inside the plastic were chicken heads and other bird bones. We burned them. These had been placed on the roof over the main entrance to ECONSULT and over the door leading into our computer programming classroom. The night watchman, our interpreter, informed us that this had been done to produce the strongest effect at the door to keep people out of ECONSULT.

By that same afternoon, people started dropping by and calling about when our next class would start. Within a couple of weeks, our programming classes were back to normal. I had to continue sweeping the roof and found new spells there on an almost daily basis, but none had any impact on our business. This did not mean that my enemy was not determined. The rainy season was upon us, and one day when it was raining very heavily, a leak developed in my office, falling from the ceiling onto my computer. I had to move my desk to protect it. I got a ladder and climbed onto the roof, taking a spare clay tile with me. I found the broken tile and removed it. Before I had a chance to replace it, my eyes caught sight of something inside the roof. It was another chicken head. My opponent had climbed on top of my office, removed a tile, and inserted a chicken-head package inside my roof. Had he noticed that he had broken a tile, I might not have caught him. After a few weeks, the frequency of spells dropped until they stopped.

ON USING CANDY FOR MONEY AT THE SUPERMARKET

Whenever I went to the supermarket, I was never given the correct change. If the difference between the cash I paid and the bill was less than a dollar, instead of giving me my exact change, the cashier would take a sack of individually wrapped candies, pour a few onto the counter, and call it even. All Brazilians played along and accepted this. I did not. This irritated me. The cashiers did not even ask me if I liked that candy or if there was another candy I might like better. They never asked me if I accepted the exchange between candies and money. I was supposed to accept it and move on. The supermarket was a big business that should have a way to give exact change. It could easily serve several thousand customers a day. It made money from that play.

The next day after one of these occurrences, I bought a bag of individually wrapped candies elsewhere and filled all my pockets with them. I reentered the supermarket and grabbed a can of Coca-Cola. I headed to the line to await my turn with the cashier. When it was my turn, I placed the Coke on the table, pulled a few candies from my pockets and threw them onto the counter, grabbed my Coke, and started to walk out. The checkout lady at first did not understand and was speechless. Then she understood and yelled. Everyone started to watch us. She asked me what I was doing. I told her that I had just paid for my Coke with candies. She said that was ridiculous because candies were not money. By this time a couple of security guards were watching, and people behind me were joining in by defending the supermarket and swinging their fists at me for delaying their checkout. I asked the cashier if she gave out candies when she did not have exact change. She agreed that she did. I told her that candies were in fact money then because the store itself was using them as such. Now the upper management had arrived, and I repeated my argument to him. He seemed perplexed and did not know what to do. I was enjoying myself because I was making sense to everyone except my fellow shoppers.

Finally, to avoid being hung from the rafters by other supermarket shoppers, I grabbed my candies and paid in cash. The cashier had to borrow money from a neighboring cashier so that she could return my exact change. She did not want to push her luck and give me candies in lieu of change. I left. My job was done. I had needed to do that. It was so easy to create good drama.

ANOTHER WHITE TABLE VISIT

Katia and I were feeling down. We decided that we needed a visit with Preto Velho. We made the appointment and bought the required vile cigar and a flask of firewater.

After Preto Velho appeared, we lit the cigar for him and opened the flask of firewater. After he'd had a couple of draws on the cigar and a couple of sips of firewater, he was open for business. He asked me what I wanted. I told him that I missed my family and wondered what they were doing. He told me to select one person and focus my mind on that person. I thought about my mother. After a few seconds of silence, he told me that she was in the kitchen and might have a headache. She was seated at a funny table (our table had a large lazy Susan in the middle), and she had a cup of something, which she was stirring with a spoon (Mom was a tea drinker). With her other hand, she was holding her head, like it hurt. Preto Velho said he was leaving there because he could not understand anything the people said. He said they talked funny (English).

Preto Velho volunteered to me that in an earlier life, I had been a farmer in Ireland. Well, I had not seen that coming. As usual, he had to suddenly dart off to attend to another caller, but I felt better.

GARBAGE ON OUR LOT

When we bought our house, we also bought the lot next to our house. We thought we might need more room later. For the moment, we left it as it was: an empty lot without the protection of an outer wall.

The city block catty-cornered from our house was full of houses. On the far corner, located on a double lot, was a large two-story house owned by one of Natal's more prominent lawyers. As I sat on my front porch, I saw a maid exit this house with a plastic bag of garbage balanced on her head. I had a bad feeling about this. I watched as she walked in the direction of our property and could see the plastic bag bobbing up and down over our outside wall. She disappeared, only to reappear a few seconds later without the garbage on her head.

I knew what she had done. I ran outside and saw her garbage on our empty lot. I followed her back to her house, knocked on the door, and asked the occupants to remove their garbage from my lot. They did not.

After a reasonable time, I returned and knocked on their door again. The maid opened it only a couple of inches and said nothing before closing the door. I did the "oh de casa" thing and clapped loudly three times. This alerted anyone inside that I was outside and was addressing them. I told them to remove their garbage from my lot, or all my future garbage would be slung over their eight-foot wall onto their beautiful property. I told them they could count on this happening if they did not remove every item of garbage they had deposited on my lot. Within a half hour, the maid made another trip to our lot and removed the garbage. I had learned early that in Brazil, if you did not stand up for yourself, people would walk all over you, except if they carried machine guns. Then you should let them walk all over you.

MY NEIGHBOR WITH A MACHINE GUN

We had sold our original house and rented another that was located in a densely settled area, even the streets were narrow. It was hot, very hot, during the dry season. At the entrance to our garage, a nice shade tree had grown in a hole placed for that purpose in the sidewalk located just outside the wall. This allowed me to come home for lunch and park my car in the nice shade of our tree—except suddenly, I could not. Our neighbor also came home for lunch, and he did not have a nice shade tree because he had not planted a nice shade tree. He used my shade tree—he parked his car in my shade. I considered this to be the other side of rude.

I had learned to confront people rather than accept being stepped on or used by other people. When I shared my plans for revenge with Katia, she told me to leave him alone and park in the sun. She had a bad feeling concerning the nature of our neighbor. I always listened to her intuition. She was never wrong.

One day, I noticed our neighbor arriving. He got out of his car and collected his things before entering his house. His things included a briefcase and a machine gun. Now I understood Katia's warning. We started to secretly watch his comings and goings. Every time he entered or left his car, he always had a machine gun under one arm.

A few months later, this neighbor had some issues with the law. We learned that he owned and operated a strip club in Natal and that he had put a hit on some people who had upset him. Unfortunately for him, the people he had put a hit on were themselves very well connected. Our neighbor left to spend several years in prison. Now I had my shade tree back, and I had gotten it back without confronting my neighbor.

EMPLOYEE PROBLEMS

Every business eventually has employee problems. I had a young man who did my accounting for me. I do not remember why, but it was necessary to separate him from the business. He did not accept the news well. He told me that I would be sorry and left.

When I arrived at ECONSULT at 8:00 a.m. the next morning, I had three serious gentlemen waiting for me. They were there to check my books to determine if they were up-to-date and in order. One man represented the city of Natal, another represented the county of Natal, and the third represented the state of Rio Grande do Norte.

I turned over all our books for them to review. After a suitable time, they all found their way to my office, where one by one, they announced that they had found grievous errors in my bookkeeping; however, each man said that if I just gave him a fifth of Johnnie Walker, he would not bother me.

If I agreed to their demands, they could return another time and find the same grievous errors. I told them that I did not pay bribes. If my books were wrong, I wanted them to show me, and I would correct them. If there was a fine for me to pay, they could give me an invoice, and I would pay it. I told them that I wanted my books to be impeccable. They all left and never returned. I found it interesting that by supposed chance, they each had asked for a fifth of Johnnie Walker whiskey. I also found it to be an interesting coincidence that they all had arrived the day after I fired my bookkeeper.

CHASED BY A MOTOR SCOOTER

I always returned home for lunch. On this rainy day I turned onto the main street that passed in view of my home. The street was very wide because it was a main street, and it had no effective ditches. People used the flat ditch areas like they were extra lanes. There could have been three lanes in each direction, except it was a dirt road with many potholes. People did not drive in a straight line; sometimes they could swerve suddenly to dodge a pothole. For this reason it was not wise to pass close to any vehicle. It was always best to give other vehicles a wide berth.

I was traveling at a normal speed when my front right tire hit a pothole filled with water. A jet of water shot out from the hole. By coincidence a man on a scooter was feet away from that tire and was absolutely covered in water and almost knocked from his scooter. He blamed me. In fact, once he regained control of his scooter, he started chasing me while waving his fist at me. With an angry man chasing me, I could not turn toward my house. I had to lose him first, so I drove on. I was in the city, so I had to be careful, or I would hit another vehicle or a pedestrian. With those limitations, I could not lose him. In fact, he was tiny, but he could pick me out of a crowd several blocks away because I was driving a VW van. It stuck out.

All I could do was avoid streets with streetlights and keep driving, hoping that he would tire. When I looked in my rearview mirror and no longer saw an angry little wet man on a scooter shaking his fist at me, I turned my van toward home.

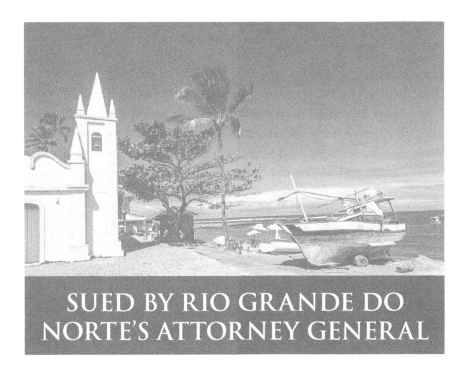

SUED BY RIO GRANDE DO NORTE'S ATTORNEY GENERAL

I met an American man who was working on a project that involved modernizing Natal's fishing fleet. He wanted to replace the homemade floating hunks of wood that used a homemade sail with diesel-and-sail-powered boats. I read his plan for the project, and on paper, it looked good. That's when he asked me to be on the organization's board of directors. I talked it over with Katia, and she saw no problem. It did not pay anything and did not appear to require any work. I signed.

A few weeks later, I received a personal visit from Rio Grande do Norte's attorney general. The state was suing me for some irregularity on the project. I tried reaching my new friend, but he was nowhere to be found. He had flown the coop.

The attorney general was a character. He was young, maybe thirty years old. That told me that he was very well connected and had powerful friends. He was dressed impeccably in a suit that looked very expensive. His haircut looked like it had been done that morning. When he visited me—and I could not figure out why he was visiting me—he insisted on sitting in my chair and addressing me across my desk while I sat in one of the guest chairs. I did not like this guy.

Finally, everything fell into place. He had just bought a computer from someone else, and he needed a five-inch floppy drive. He had heard that I had one. In those days it was very difficult to get many computer components, including printers and drives. I confirmed that I did have one. He said that if I would give it to him and install it for free, he could make the lawsuit disappear. I took the deal. I did not want this man in my store or at my desk again. This was the only time I paid a bribe.

MY FRIEND KILLED OUTSIDE A NIGHTCLUB

Katia's family was close friends with another family who had a daughter near my age. This family was very prominent and was friends with everyone, even those with politically opposing views. Most people had to pick a political party and hope that it won in the elections, but this family did not have to worry. They had friends in all political parties. It did not matter who won the election; the members of this family would still have excellent jobs and could obtain financial loans. Their lives would continue normally, and they would flourish. There were not many people in the state who could enjoy those benefits.

The daughter had also been my boss at CEPA. Besides working at CEPA, she owned and operated a large ranch. She had obtained a significant financial loan at a negative interest rate (i.e., monthly inflation was higher than the interest rate). On paper such loans were available to everyone, but in reality they were not. Only a select few had any chance of obtaining one, and she was one of them. She had barbecues with the governor and maybe with Brazil's president and vice president. She was connected.

She and her husband liked to unwind from their busy schedules. One night they went to a nightclub, the same one where Katia and I had met. She and her husband were drinking. As they drank more and more, they danced more aggressively and sometimes bumped into other people. Because people were full of drink, words were likely exchanged. She and her husband probably did not slow down or apologize because they never had to do that. They likely annoyed many people who just wanted a nice night of dancing.

Upon leaving the club, they entered their car and started for home. Their vehicle was blocked by another vehicle. Men exited with guns, and both the woman and her husband were shot several times. She died at the scene, and her husband needed years of physical therapy, after which he could walk only with the use of a cane. Her murder and his assault were crimes that were never solved.

In Brazil, you must assume that everyone is armed and is capable of seeking revenge. They are, and they do.

MOVE TO SÃO PAULO

I tired of not selling computers. I tired of not repairing computers. I tired of trying to push water uphill. I worked long hours and never saw my children except on weekends. I always left for work before they awoke and arrived home at night after they had gone to sleep. I had to deal with more and more employee problems, and I was aware of how fast we could go from just surviving financially to dire straits. I had added the computer enterprise to ECONSULT to lend support to my consulting. Now I had no time for consulting because I was busy solving computer-related problems.

I confided this to my friend who managed the Algodoeira São Miguel. He invited me to a party that he was hosting. It involved whiskey and sandwiches. David had also been invited. The manager told me that in attendance would be a man from a company in São Paulo that created agricultural software. There would be other men present as well, but I no longer remember who they were. I met at least a dozen men at the party and enjoyed our conversations. I appreciated the Algodoeira manager trying to help me, but I did not see myself getting any closer to a solution that would allow me to move to a market more open to my services.

I saw no future in Natal. It was a region where one's future depended entirely on who your friends were. Your individual capability played no part. Natal was a city of nearly a million people, but it was owned and operated by one or two dozen families. If you were not connected to those families, you were on your own. I was on my own.

My experience was that when one person started to get ahead, there were one hundred people there to drag him back down again. There were many ways this could happen. Rumors could be spread, and they spread very quickly. Many times, deals that were certain one day were impossible the next day because someone had reached out and killed the deal in some unknown way. The teachers I hired were stealing my class materials and starting their own classes on the side. Let's also not forget the role of "spells." I was exhausted from fighting these wars. I just wanted to go back to my consulting without anyone interfering in my results.

Katia and I were uncomfortable. We did not want to leave our friends and family in Natal, but we knew that there was no future for us there. We decided to say hello to our friend Preto Velho. We always felt better after talking with him.

As usual, he was not serious until he'd had his puffs on the cigar and a few swallows of firewater. I explained our situation to him. He asked if I had met any new people recently. I told him that I had. He asked me to concentrate on one of them. I made my concentration face for a few seconds. He said, "No, not that one, the one behind him."

I moved my focus to the man I had met from São Paulo. Preto Velho confirmed, "Yes, that's the man! He will offer you a job in São Paulo within a few days."

I must admit that I felt better, but I did not believe it. Four days later, the man from São Paulo offered me a job. As part of the offer, the firm would pay for our move in its entirety. I accepted.

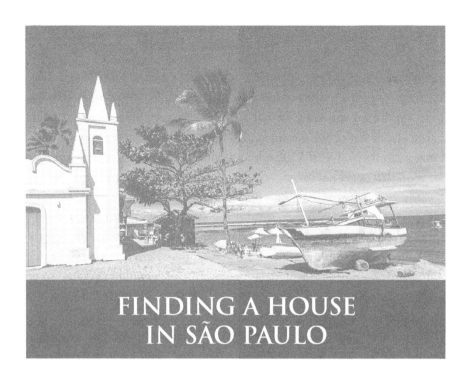

FINDING A HOUSE IN SÃO PAULO

M y first task was to fly to São Paulo and find a house

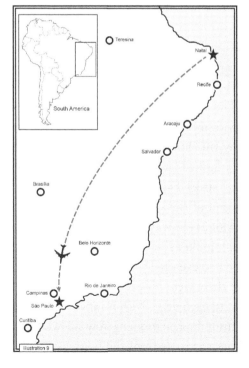

My new boss gave me directions to his business, and I took a taxi there. He showed me my new desk and a map of São Paulo. It was so much larger than anything I was used to that I almost wet my pants when I thought of the task at hand. Within a couple of days, I had to find a suitable place for my family to live; it had to be a home that we could afford and in a location that would allow me to get to work in a reasonable time.

My boss showed me the communities that were too expensive for my salary. He ran his finger around the map, showing me how far out I should start looking because the farther I went out from the city center, the more rents would drop. Anywhere closer than he was advising, I could not afford the housing. I made a mental note of how far out I had to go on the subway before I could get off to look for housing. I thanked him and started my journey.

I found one leg of the subway and rode way out before stepping off. It looked like it could rain at any moment. Great! Now what should I do? I walked and walked, looking for houses for rent. I wrote down numbers, and then I searched for a phone booth. Once I found a phone booth, it was usually impossible to reach anyone at the numbers listed on the signs, and if I did, the homes were not available until much later, and that did not help me.

I kept looking. I stopped at another subway station and started to make a small circle around it, looking for "for rent" signs. I decided I would keep making the circle larger until I found a house. After an hour or two, I found one. The sign pointed down a steep dead-end street, an offshoot from a street with moderate traffic. That was perfect because there would be no traffic by this house. There were many cobblestones missing from the street. They looked like they had been washed out during heavy rains.

The stubby street had only a few nice houses on it before it dead-ended. At the end of the street was a six-foot adobe wall built across the street, with a door in the wall. Behind that door was a completely different neighborhood. It looked like it had probably started out as a neighborhood of squatters and had eventually become legitimized. There was no vehicular traffic because there were no streets, only narrow sidewalks. Houses were built right next to each other with no space in between. A few feet outside one resident's front door, across a narrow sidewalk, was someone else's front door. Sometimes the roofs from the houses on the two sides of the street nearly touched each other. Since these poor people had no cars, they did not need to leave space for cars to occupy. There was no wasted space.

The last house on the right before the wall was the house for rent, and the owners lived next door. The house was new. It had a two-car garage surrounded by thick metal bars. The enclosed garage area could become a nice, safe place for the kids to play. The house was narrow, maybe less than fifteen feet wide. Inside, the living room and the dining room had nice hardwood floors. Between the living and dining rooms was a half bathroom. Past the dining room was a kitchen. It was not large, nor was it small. Behind the kitchen was a space for the maid to live and to wash clothes by hand. Between the kitchen and the maid's room was an open space designed for clotheslines to dry clothes. Upstairs there were three bedrooms and a bathroom with a shower. Outside the room that would be our bedroom was a very large open patio from which we could view São Paulo. All floors were tile, except for the living and dining rooms.

I wasted no time. I had the owner fill in the contract, and I signed it and paid the first month's rent. I felt good. I took the subway back to the airport and caught a flight home yet that night. I was so relieved. To find a house in such a large city was a monumental task.

Christmas was approaching. In early December I announced to our employees that we were closing shop. I told them they would be paid through the end of December. I sold off the inventory and assets. I was happy but nervous.

We packed our house's belongings into a truck. The truck driver agreed to meet us at our hotel in São Paulo at 10:00 a.m. on December 28. We flew out of Natal on December 27. Kevin was days from turning four years old, Nicholas was days from being three years old, and Christianne was three months shy of one year old. It was not an easy trip.

We arrived at the hotel and spent the night. The next morning, we checked out and waited in the lobby. At 10:00 a.m. the truck driver walked in. I had never believed he would meet me on time. Given my experience of people always being late in Latin America, I would have been surprised if he had even showed up, much less showed up on time. This was a good beginning to a new life.

I caught a taxi and gave the driver our address. The truck driver followed us. It took a while for us to reach the house because we had to make sure that we did not become separated in the heavy traffic or at traffic lights. When we reached the final turnoff to the dead-end street, I had the taxi stop, and I showed the house to the truck driver. He would have to back down the street. I grabbed our suitcases and led Katia and the kids to the house while the movers started unloading the boxes. I had to stand at the door and direct them to the correct rooms.

Soon, the movers were gone. I took Katia around the house and showed it to her. I had hoped she would be as excited as I was. She was not. She was not unhappy, but she was not as excited as I was. I hoped that time would change her opinion.

I started work right away. My first assignment was to do a quality check on large agricultural investment software that had been created to aid large businesses in making huge investments in agriculture. For example, VW had a farm with hundreds of thousands of acres. Banco Safra had a ranch with over three hundred thousand acres. And there were many more.

The government was sponsoring such projects. Brazil had a program wherein if a company bought undeveloped land, the company either could pay millions of dollars in its normal taxes or could pay no taxes and invest in the company's ranch the money that it otherwise would have paid to the government in taxes. So instead of paying taxes, the companies owned land. This was a program that entirely favored the rich. There were dozens of such projects around Brazil. Each project was worth millions of dollars. This software program targeted these companies.

My boss had brought a young Englishman to Brazil for a year to develop the software program. The business had then deemed the program ready and sent him home. The program was phenomenal. When I arrived, I asked to check it for defects before it was released for sale. My boss reluctantly allowed me to test it. I told him that it was an expensive program destined for only a few companies. If one such company bought the program and discovered problems, it would share that knowledge with everyone else that might purchase the program, and our company would make no more software sales.

On the first day of testing, I discovered a major problem. Under a very specific set of circumstances, the program blew up and gave ridiculous answers. After a week's work, I isolated the problem to within two or three lines of programming. The lines seemed perfect, yet they did not yield the correct answer.

The company brought the Englishman back to São Paulo to solve the problem. In the end, it was an idiosyncrasy in the BASIC language in Apple computers. One line placed in the middle of the program had created a problem. Once the line was moved to within the first few lines of programming, the program worked perfectly. That was all. I just copied, deleted, and pasted one line of code. It probably cost the company ten thousand dollars to discover and correct the problem. Now the company had an expensive product ready to market.

I immediately learned that working in this office was not going to be easy. There was a female computer programmer and a female office manager. They were both young and loved to gossip. One problem was that the office manager was the mistress of one of the partners. The two ladies spent much time together gossiping. I was in the next cubicle, trying to work. Their gossip included things about me. I never understood whether they did not care that I heard what they were saying or they believed that I could not hear them. It was unpleasant, and I did not think that complaining would help.

TWO WEEKS AT A CONVENTION IN RIO DE JANEIRO

A two-week agriculture convention was coming up at the convention center in Rio de Janeiro. Every business even remotely connected to agriculture had bought space at the convention center, including my bosses. People involved in agriculture from all over Brazil would be there. It would be an excellent chance for our firm to snag some business. The company elected to send me and contracted another person from outside the firm to go with me.

My bosses had friends who lived in Copacabana in a two-bedroom apartment. The friends were willing to sublet the apartment to my bosses during those two weeks because they would be out of the country. I was not happy about being away from home for two weeks, but I had no choice. I was excited, though, about the contacts I might make and the things I might see.

The convention center opened at 10:00 a.m. and closed at 9:00 p.m., and we had to be there during the entire fourteen days. The routine quickly became monotonous. Only a few people stopped at our booth. We tried to sell them our expensive computer program, but 99.9 percent of the visitors did not need it. I tried to sell them consulting with cost-control software, which would be developed by our company, but no one was interested.

People did not see the need for these services. They likely thought they were capable of operating their farms or ranches without outside help.

My partner and I traded off maintaining the booth. One sat and tried not to look bored while the other explored the other booths. There were many interesting booths. I learned that our company was not prepared for this expo. We needed to have more things ready to show. We needed posters showing the benefits of rotational grazing or showing the advantages of knowing the cost of production. The producers needed ways to make forecasts of their costs and revenues. I was frustrated. I felt the company must have become aware of the expo just a couple of days before deciding to send us there. No planning or preparation had been done.

I obtained a few business cards, but no one showed any viable interest in our services. At the end, I only wanted to go home. I did not look forward to reporting to the two owners that our efforts had not been fruitful. I had the feeling that they would think our lack of sales was my fault.

LEAVING THE COMPANY

When I returned to the office and had to listen to gossip again and feel like I was blamed for the lack of sales, I found myself unhappy in my work. I felt that this company was similar to ECONSULT in that it had many problems, one of which was that it had no sales force. I think that was why they hired me—to sell things—but if they would have clearly stated that to me, I could have told them that it was not going to happen. I did not have sales expertise. I detected that they did not have nearly enough revenue to sustain their workforce, and this was the reason they were on edge all the time.

One of the bosses ran around the world doing consulting for the World Bank. After being gone for a month, he would drop into the office to finish his reports. He worked day and night. He could have lived a good life had he used his income just for himself. Instead, he sank his earnings into a company that had no chance at generating revenue because no one was making sales or supervising the workforce.

The other boss spent most of his time on his farms, supervising work there. This left our office unsupervised and in the hands of the office manager. I felt the company was doomed to fail for nearly the same reasons that ECONSULT had failed as a computer-based company.

One day, a person I had met at the expo asked me if I would be interested in working with him. Most ranch owners lived in the city and had other business interests. They looked at their ranches like an occasional hobby. They would only visit their ranches from time to time and had full-time in-house ranch managers to take responsibility for the day-to-day activities. My client was not like this. He was a rancher who lived on one of his five ranches, and he made his living from ranching. I decided to leave the software company and work directly with this rancher.

He wanted me to look at his pastures and his operation to give him advice

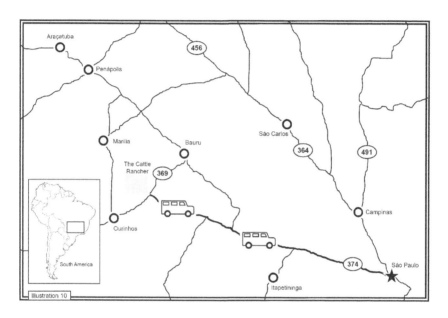

I suggested he divide his large pastures into smaller paddocks and use rotational grazing. I felt that he was overgrazing all his pastures. This was bad because it caused his grass to slowly degrade. His future revenue depended on the quality of this year's pastures. I showed him where he needed to place electric fences for cross fencing. Water had to be placed at strategic locations to keep the cattle close to water yet dispersed on the grass.

I created data-collection forms that were designed to document how his cattle herd was moved from pasture to pasture and paddock to paddock. From this I could determine how hard each paddock was being grazed. Sometimes he might need to remove animals to lessen the grazing pressure on certain paddocks. As the grass entered the dry season, he needed to leave all paddocks with considerable leaf growth because root growth depended on leaf growth. Grass without a good root system was less likely to survive the dry season.

This rancher was very nice. He told me of some of the problems that he was having. There were dozens, maybe hundreds, of landless peasants encamped along the road that passed by his pasture and into town. During the dry season, they would often try to set his pasture on fire, although usually not much damage was done because he had overgrazed his grass, leaving nothing to burn. They would also cut his fences, allowing his cattle to escape onto the highway. They were trying to drive him off the land so that they could split it up among themselves.

Once when I went to visit him, I was told that he was gone. I reminded his wife that we had made an appointment for that day. His wife said, "Yes, but the peasants tried to kill him the other day when he was driving into town. They fired several shots at his pickup. They left holes in his rear window." She assured me that he would be back in a few days, after things cooled down.

THE CHICKEN RANCH

One person with whom I had spoken at the Rio expo was part-owner of an operation that produced eggs, among many other things. He had given me his card. I called him and asked for a meeting to explain what I could do for him. He invited me to his ranch so that I could meet the other three owners and visit their business enterprises

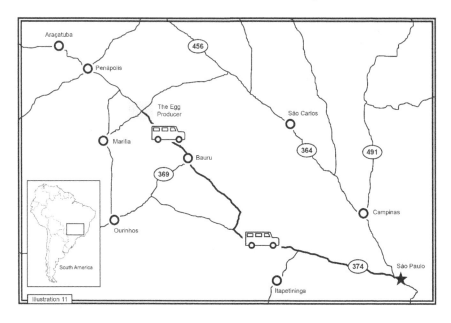

Their story was complicated. Two sisters had gained ownership of this large operation when their parents died tragically in an automotive accident. One sister, age twenty-four and single at the time, was a veterinarian. The other sister, age twenty-two and single at the time, was an economist. Together they owned 160,000 laying hens, 2,000 head of cattle, a small system of pork production, a transportation company, and five ranches producing coffee, corn, black beans, and soybeans. Within a few months of receiving these assets, they had found husbands. One had been a veterinary product salesman, and the other had been a beer salesman.

As I made the four-hour trip, I tried to think of how I could help them. Without seeing the group of enterprises, I had no idea. But I knew that they would want me to explain how I could help them within minutes of arriving. I would have to deflect that question until later.

When I arrived, they led me into their office. It was a very large room with no partitions, and it housed six to eight office employees. I learned they had nearly two hundred employees total. They offered me a tiny cup of sweet coffee and each took one themselves. Before I could finish my coffee, they were already asking how I could help them. I asked if I could refrain from answering that question until I had seen the entire operation.

The egg operation consisted of eighteen batches of 10,000 birds, 160,000 of which were laying eggs. Every batch of hens arrived together as chicks and stayed together until they were shipped out to soup factories. I saw that the business had no individual production records. What production records they had were always for the combined 160,000 laying hens. I made a mental note of this.

They used their transportation company to distribute eggs to their egg buyers and to bring back truckloads of corn that had been imported from the US. I watched a couple of men grinding corn with other ingredients to make feed rations. They had a total of five or six different feed rations to accommodate all the ages of chicks and chickens.

The cattle operation was in good condition. Their pastures were not overgrazed, and their herdsmen knew what they were doing. The same went for the business's coffee and grain production.

We met again late in the afternoon. My plan included the following: count the eggs by size for each of the sixteen lots and create a spreadsheet that would allow the owners to predict how much cash and revenue would be generated month by month during the life of each lot. To do this, I would need to create data-collection forms and train people to fill them out correctly.

The chickens laid eggs that were classified into one of five or six different sizes, ranging from small to jumbo. Each egg size had its own price. To predict cash flow, we had to predict the distribution of egg sizes. Fortunately, egg sizes depended on the hens' age and breed and were predictable. Since each of the eighteen lots had a different hen age, each lot had a different expected distribution of egg size and a different expected flow of revenue.

The people on the farm who received the eggs from each lot were overwhelmeded when the eggs started coming in from all eighteen lots. Usually, they just piled them up and processed them as they came in. Now they were being asked to pile the eggs from each of the sixteen lots separately. They were more than unhappy—they were near revolt—but with time they learned to manage easily enough.

These data were recorded into a spreadsheet each day. After I showed the office manager how it worked, he could easily spot an error in data if, for example, one case of eggs that belonged to lot 16 was placed in lot 10 by mistake. That ability to quickly spot an inconsistency was a management tool.

One week the office manager noticed another inconsistency. One lot of hens was producing only half what they normally produced—of all egg sizes. He brought this to the attention of the veterinarian. She visited the building to investigate and learned that the chickens were producing half their eggs without a shell, and these eggs broke as they hit the wire mesh at the bottom of the cages. The egg collectors had never mentioned this because no one had ever cared when it happened before. It apparently had happened many times before.

The veterinarian collected feed samples as the feed left the feed-grinding operation and again when it arrived at the affected hens' housing. This particular lot of hens was housed about a mile away from the main operation. The dirt road that connected the two locations was filled with potholes. The veterinarian found that the feed was fine when it left the feed mill but was deficient in micronutrients when it arrived at the hens' housing. She concluded that the rough road shook all the micronutrients to the bottom of the feed wagon, and the augers used to empty the wagon did not pick them up. She started having the micronutrients added at the destination, and the problem was solved.

Each day that the hens laid eggs without shells cost the company from 250 to 500 dollars or more. Before, the loss of 5,000 eggs out of 160,000 eggs from all the lots had not been detected. But now the loss of 5,000 eggs out of 10,000 eggs from that specific lot demanded attention. It was possible that the company had suffered tens of thousands of dollars in undetected losses over the years. This simple system was a powerful management tool.

Over the next couple of weeks, I worked on the spreadsheet that would estimate the feed needed and the cost of producing each lot of chickens. If printed, the spreadsheet would have taken nearly twenty pages. It was very sophisticated. I was proud of myself.

THE GERMAN PIG FARMER

I do not remember how, but I met a young German pig farmer whose name was Gerd. He was in his early forties, and he was handsome, rich, and friendly. He worked as an oil broker during the week, and on the weekend, he was a gentleman farmer. His father was on the board of directors of some very large bank in Germany. The father was very wealthy, and Gerd had many valuable investments in Germany. He confided in me that he lived off his German income and played with what he earned as an oil broker.

Gerd was married to a beautiful younger woman with dark skin and jet-black hair. She was allegedly an Indian princess. They had several children together and lived on an entire city block in São Paulo, where real estate was very expensive. As a couple they seemed to be very dedicated to each other. I enjoyed watching them together.

Gerd had a farm about three or four hours outside of São Paulo. On his farm, he had a beautiful house with many bedrooms, a beautiful kitchen, and an enviable living room, which featured a cozy fireplace and a sunken area whose built-in sofa could accommodate a dozen or more people. Gerd liked to host parties for his friends on his farm. He was very sociable.

Gerd showed me his livestock operation

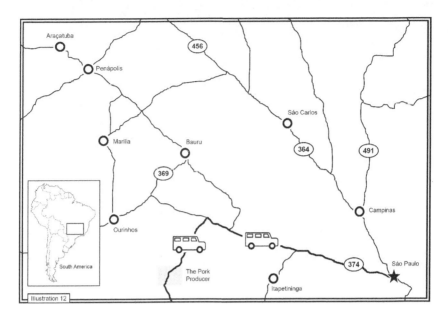

Illustration 12

He had a couple hundred cows and a capable herd manager, although his pastures were overgrazed. His main enterprise was not cattle but pork production. He had already built several housing units for his sows. As soon as the sows' piglets were weaned, they were moved on to another housing unit, where there were introduced to a growing and fattening regime.

Each month, he added more production units. I do not remember how many sows he had at the time, but his production could no longer be considered small. His goal was to become a large producer of pork. He was increasing his production capacity slowly but steadily. Gerd needed to keep track of how much money he was investing in each housing unit, and he needed to know his costs of production. I recommended that he start collecting data so that he could manage his investment. I would create the forms needed.

Gerd was a member of an exclusive local cooperative. Not all farmers could join. He explained to me that only farmers of German, Swiss, or Austrian blood could join. The group excluded Brazilians because, he said, they had difficulty keeping their word. When the co-op agreed to invest money in a new structure for the co-op, members agreed to put up money. They had discovered that Brazilians often made promises they could not or would not keep, whereas the German farmers did exactly what they had said they would do. For the co-op to make progress, they could not waste time with people who did not keep their word; therefore, Brazilians were excluded.

Once in São Paulo, Gerd invited me to lunch at a very expensive restaurant. I could not imagine why we needed to meet there. Worse, I had to wear a suit. I went because he asked me, and I could not refuse him anything. I met him and his wife at the restaurant. Gerd was always dressed in a suit, but previously I had seen his wife only at the farm, where she dressed for lazy weekends. On this day, she was dressed in an exquisite dress and indeed looked like an Indian princess.

We were seated at a table in the middle of the restaurant. I did not feel at ease or at home. As we received our menus, Gerd and his wife were busy identifying who else was at the restaurant. Gerd pointed out three or four men who were in charge of multimillion-dollar agricultural projects. Some I had read about in the newspapers. The restaurant had a balcony, and Gerd kept squinting at a table located in its far corner. Finally, he recognized the man as the head of a company starting an agricultural project worth more than ten million dollars. He was thrilled. He had not yet opened the menu.

In the end, we never talked about business. I figured out that I was an adornment that they needed. They had not wanted people to think that they had come to the restaurant only to see who else was there, so they had brought me. If anyone asked them, they could say they were meeting with their agricultural consultant.

I found just watching Gerd and his wife in that scenario interesting. I could see that Gerd wanted to be that man at the corner table on the balcony more than anything in the world. Each time he added a new pork production facility, he was one step closer to realizing that dream.

A few months after we started working together, Gerd invited me to his house for relaxation. He told me that they were killing a yearling or two and having a barbecue. They had invited a few friends over. I thought that was an excellent idea. Cold beer and barbecue always made for an enjoyable afternoon.

When the day came, I was sitting in a chair next to Gerd, drinking beer as the yearlings roasted on a grill. The smell was enticing. Then the people started arriving. Not two or three, but twenty or thirty. They were all German, Swiss, or Austrian farmers. It seemed Gerd had invited all the members of the cooperative. Most were in their sixties or even seventies. One elder gentleman rode in on his horse. He wore a riding outfit with flared legs between the hips and knees and black riding boots up to his knees. As he walked toward us, he slapped his riding stick against his boots. I looked at his head, expecting to see an SS hat with a Nazi insignia.

A lot of German was being spoken. I felt out of place and outnumbered. Many people spoke Portuguese, but it was the first time I had heard Portuguese spoken with a German accent.

Later in the evening, after darkness had crept in, the conversation turned to World War II. I stayed silent. They talked about what had happened to their fathers and uncles. Some had been killed in the war. Others had been wounded, and still others had psychological problems. Some of the older gentlemen had themselves been in the war. They were mostly silent.

Many Germans had decided to leave Germany and go to Brazil toward the end of the war. Some fled from the Allies to escape war crime allegations. Anyway, I stayed silent, but even if I felt out of place, I did not want to leave either. I wanted to hear whatever they had to say. It was a unique opportunity to learn about people by listening to their stories while they were completely at ease. Most of them did not realize that I was an American. These stories were worth the long trip to the farm.

SURPRISE—ALL PRICES ARE FROZEN

In spite of all the government's best efforts to contain inflation, it steadily increased. If my memory is correct, it was between 15 and 18 percent each month. At 15 percent, goods costing 100 dollars at the beginning of the year would cost 535 dollars at the end of the year. The government was worried because working people, unemployed people, and retired people were being devastated by the effects of high inflation. Businesses were fine because they just raised their prices as inflation increased. Working and retired people could not just raise their wages and pensions.

One problem with such inflation was the difficulty in making large purchases. Let us assume you wanted to buy a new car. You visited one store, and its price was 24,000 dollars. You wanted to make sure you had the best deal out there, so you visited another store but found that it was charging 28,000 dollars for the same car. You ran back to the first store, only to find that while you were out searching for a better deal, it had raised its price to 32,000 dollars. The problem was that when you were making a purchase, you had to be in tune with the market so that when the market gave you a fantastic price, you could recognize it and jump on it. If you waited, you would lose. Yet this mentality helped feed more inflation: buy now no matter what the price; otherwise, the purchase might be more expensive later.

One fine morning, Brazil awoke to the news that the government had declared all prices frozen as of three days previous. Whatever price businesses were charging three days ago was the maximum price they could charge now and in the future. Some people had just adjusted their prices for inflation a few days earlier. They would be in good stead because they always added a little extra to the price for unexpected inflation. The problem was that the retailers that had not adjusted their prices for a while and that had been preparing to increase their prices, perhaps by as much as 25 to 35 percent, were caught in an unsustainable situation. Their selling price was frozen at a level insufficient to cover their costs, much less give any profit.

Those businesses caught with an unfavorable price had little choice. Their only protection against bankruptcy was to stop the manufacture or marketing of their items. Each week there was more and more empty shelf space in stores. Shopping lists for the supermarket were useless. You bought what you found on the shelves. If you found toilet paper, even if you did not need it, you bought it because it might disappear and not reappear for months. Within weeks, at least half of grocery store shelf space was empty.

It was true that common workers saw benefits from this government action. Their money went further than it ever had before. Their standard of living improved. They had never been able to eat meat before, so living in an economy now without meat did not impact them.

The biggest fight broke out between the government and ranchers. Beef was a product that happened to have a very low price when the government froze prices. Ranchers could not sell their beef without incurring a financial loss. They were steadfast against this. For the most part, beef disappeared from the butchers' shelves and was not replaced. People shifted consumption from beef to pork and chicken until those meats too became difficult to find.

During a six-month period, we never ate beef or pork. We sometimes found chicken, but more often we found only eggs, and there were times when we could not find even eggs, but we were fine. Rice and beans were a meal, or when we had eggs, rice and eggs were a complete meal. We did not suffer. We always had enough to eat.

In Brazil, farmers and ranchers depended completely on loans from the government. These loans always involved large piles of paperwork, but even so, they were worth the effort. These loans were offered with negative interest rates. For example, if inflation was 535 percent annually, a loan might have a 20 percent interest rate. That was a huge advantage for the rancher. Large agricultural enterprises often had one employee whose job it was to obtain all these loans that the company could possibly obtain. The loans were not just critical to ranchers; they were absolutely a must. No farmer could survive if he were forced to go to a regular bank for a commercial loan and pay interest rates greater than the rate of inflation.

The government had grown tired of these ungrateful ranchers refusing to cooperate. They decreed that ranchers must start selling their herds of fat cattle, or they would be banned for life from receiving any more government loans. This was harsh, but the ranchers defiantly shook their fists and shouted insults at the government. The government was not happy.

After a couple of weeks, the government raised the stakes. It decreed that for every rancher who failed to deliver beef animals of appropriate weight to the butcher, the government would have a representative fly over the herds in a helicopter to determine if the animals were at marketable weight; if so, the representative would radio to a fleet of cattle trucks waiting nearby. Those cattle trucks would have a driver, a helper, and two to four soldiers with machine guns. They would enter the ranch and load the cattle. The rancher would not be paid anything for his beef animals. That was the price for defying the government.

One day during this confusion, I was making an on-site visit to my rancher client who had been shot at. We were standing in a pasture, observing the cattle and the grass's condition, when a helicopter appeared low out of nowhere. It flew slowly over the cattle, scattering the animals, and then kept flying. We looked down at the end of the driveway and saw two cattle trucks and, behind the cabs, several men with machine guns. They backed up and also moved on.

There was much pushback over the government's brutish behavior. After a couple of weeks, the government backed off. I believe that any cattle taken from owners were eventually paid for.

There were many side effects of the frozen-price escapade. One was that you could find expensive machines that were unusable because they needed a five-dollar replacement part that was unavailable. The government, in trying to control inflation, killed the economy. Eventually, the government removed price controls, and that caused a spike in inflation because everyone raised their prices just in case the government should decide to reinstate the price controls. After all, the government was unpredictable, except in its ability to make a bad situation even worse.

WINTER IN SÃO PAULO

During winters in São Paulo, the temperature could reach freezing. A couple of times, we had snow flurries. The problem was that most homes had no heating system. What was worse was that the homes were built for ventilation to make the summer heat tolerable. Rather than glass windows, we had slats. We could open or close the slats, but there were still openings through which cold air could enter the house. Above the doors, in the space between the top of the door and the ceiling, were slats that could not be shut. They were open all the time. What did this mean? It meant that if the weather was thirty-two degrees outside, it was thirty-two degrees inside too.

We had to dress the children warm, but they did not seem to mind because they were always playing. The problem for me was shower time. Yes, showers were short and to the point. No dawdling occurred.

THE POLICE AND THE THIEF

One day I was returning home, and as I was approaching my turnoff onto the dead-end street, I looked up to see a police patrol van coming toward me at high speed. Officers riding in the back and in the passenger seat were hanging out the windows, waving me to the opposite side of the road. Even the driver was waving me off the road. Confused at first, I pulled over and stopped. The driver of the police van slammed on the brakes to make the turn. The police drew their weapons, with even the driver holding a revolver in his left hand. The driver carefully entered the dead-end street while the others opened fire at something. I could see nothing to shoot at. They could only move slowly down the street because of the deep potholes. When they reached the end, three of the officers leaped from the vehicle and went through the door leading to the crowded homes behind the wall.

I followed the van down the street and pulled my car into position to park in my garage. I started speaking with the police driver. It seemed that there was a thief operating in our region. Whenever the cops started chasing him, he always headed for our dead-end street and ran through the door in the wall, disappearing into the narrow passages that linked the small houses. That was his escape strategy. He knew that any car entering the street would have to move slowly because of the potholes, giving him plenty of time to disappear. Katia said the kids had been playing in the garage. As soon as the shooting started, she had hustled them into the house and kept them in the kitchen until it stopped.

OUR WORKER AND THE STREET THIEVES

I had been contracted by a Swiss company located in the center of São Paulo. Like all offices, we had an office boy. His job was to run errands. He always carried a beat-up briefcase that contained bank deposits and office bills that needed paying. In Brazil, you never sent money in an envelope. It would never arrive. Sometimes he carried only bills to pay, and sometimes he carried ten thousand dollars in cash.

Our office was on the fourth floor in an office building. The street below, which was closed to vehicle traffic, was filled in and covered with nice, small black and white blocks that formed interesting designs. The street was at least forty feet, if not fifty feet, wide and was filled with people.

Once I was tired and decided to look out the window for a few minutes. I saw a businessman in an expensive suit carrying an expensive briefcase. I looked ten or twenty yards behind him and saw that a young boy had started to sprint toward the businessman. The businessman was still unaware he had been targeted. The boy snatched the briefcase and continued his sprint. Before the businessman knew what had happened, the boy was around the corner, where he passed the briefcase off to a confederate, who placed it inside a shopping bag and headed in a different direction. I had been told that there usually was a third member of the gang nearby with a weapon. If the thief with the stolen briefcase was attacked by anyone, the third member would shoot the person fighting with the gang member.

People were discouraged from wearing earrings, bracelets, necklaces, or even watches. These made people prime targets. Women also were discouraged from using purses with long straps because the thieves could cut the strap in a split second.

One day, our office boy was tasked with delivering ten thousand dollars to someone. Just outside our building, he was attacked. He fought and managed to get away and immediately returned to our building and came up to our office. He had a couple of cuts on his face and was scared to death, but he was safe, as was the money. That was a good day at the office.

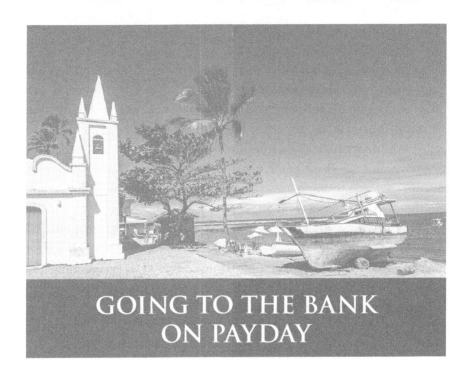

GOING TO THE BANK ON PAYDAY

In Brazil, going to the bank was an experience. Even small banks had ten to twelve cashiers working at any given time. People could not pay bills by mailing checks through the post office because their checks would be stolen. Everyone had to go to banks to make utility payments, as well as car payments and house payments. Banks were very busy places.

In São Paulo, workers typically were paid twice a month. Thieves knew this and waited for payday to rob banks. Sometimes thieves would rob one bank, walk down the street, and rob another bank, continuing the process until they were satisfied. With São Paulo traffic, it was impossible for the police to reach a bank in time to intercede in a robbery.

We never went to the bank on payday. We always managed to conduct business before or after those days.

DRIVING THE BELTWAY

The beltway was the road system around the outer edge of São Paulo. It was supposedly faster than trying to drive inside São Paulo. The beltway had four to six lanes going in each direction, with the two sides separated by a concrete barrier to keep drivers from changing their minds as to which direction they wanted to go. Sometimes traffic flowed quickly, but usually it was slow and even included periodic stoppage time.

Thieves were quick to figure out that when the cars were stopped, they were trapped. They could not move in any direction until the traffic started to flow again. Groups of thieves with machine guns strapped to their arms and with masks covering their faces would wait for traffic to stop. They then walked out into the beltway. Each thief had an assigned lane and walked from car to car, asking for donations of money and jewelry. Sometimes one group of thieves walked downstream, and another group walked upstream. The thieves even yelled at drivers down the line to have their things ready, to save the thieves' time.

Even when they had a loaded weapon in their car, gun-owning drivers did not mess with multiple machine guns. When traffic started to flow, the thieves walked to the edge of the beltway and waited for the next traffic jam. Police had no way of reaching these spots. This was in the mid-1980s, before cell phones, so people had no way of calling for help.

BANCO SAFRA

My consulting business was not growing, and inflation was hurting me. I wanted to ask my clients for more money, but because agricultural prices did not accompany inflation, I did not think it wise to ask. Farmers and ranchers had to watch their expenses. If I asked for more money, they might decide to save money by releasing me from my contract. They were already paying me what they could afford.

My solution was to look for a full-time job. While combing newspaper advertisements, I saw a wonderful opportunity with Banco Safra, the fifth-largest bank in Brazil. Even at fifth place, it was a big bank and was owned by one individual. This owner participated in the government's project to entice large businesses into investing in agricultural projects that would turn wilderness into productive enterprises. For example, if Banco Safra had a ten-million-dollar tax bill, instead of paying the government that money, it could invest the money in developing new agricultural land and, in the end, owning that land.

Banco Safra's project of this sort had started out at over 400,000 acres divided into three large parcels of land—the total acres being the equivalent of an area twenty-five miles by twenty-five miles. The land was in the state of Goias, next to Mato Grosso in the center of Brazil (Insert Illustration 13).

One parcel of more than 100,000 acres had been overrun by squatters.

Faced with the prospect of sending in a militia to reclaim the land, the bank walked away and allowed the squatters to keep the land; however, the bank doubled the number of security personnel guarding the other two land parcels. They would not give up any more land.

The company's goal was to develop 15,000 acres of land each year. The land was similar to a savanna. The team in charge of clearing the land braided three two-inch-thick cables together and connected a 200-yard-long piece between two large bulldozers. These dozers took off through the savanna about fifty to seventy-five yards apart. The cable clotheslined the trees and knocked them over. If the workers found very large trees, they went around them. These would be individually knocked down later by even larger dozers.

Once the trees were knocked down, large rubber-tired tractors with huge blades with teeth at the bottom were brought in to windrow the trees. With one tractor on each side, the workers created rows of piled trees like raked alfalfa. After the trees were windrowed, the tractors turned in the other direction and forced the rows into piles as large as was feasible. After the piles were made, they were set afire. After they were burned, they were re-piled and burned again.

As soon as the trees were cleaned up, the workers spread lime and built fences around each area. Each paddock was about eighty acres. The area was disked twice and planted to grass. The tractors operated twenty-four hours a day, in two shifts of twelve hours each, and seven days a week. It would take a couple of years before the grass could support cattle.

Each year a cattle buyer bought a few thousand cows and associated bulls to occupy the pastures that were coming into grazing. Within a few years, they would be buying five thousand head per year. After that, they would not need to buy more cows because they could add enough new cows from their own inventory of heifers. Their goal was to build the cattle population to at least fifty thousand cows.

In São Paulo, there was a group of ten or twelve employees whose job was to do the planning for this project for the next twenty years. I obtained a job with Banco Safra as the assistant manager of this group. The manager was a young man of Italian descent. He was very competent and all business. Along with me two other agricultural technical people were hired. One was a cattle specialist, and the other was a grass specialist. We had a competent team.

I had to develop a spreadsheet to project the number of cattle by age (unweaned calves, weaned calves, yearlings, etc.). I had to take into account mortality and calving rate (number of calves from one hundred cows) and other variables.

I also had to calculate the number of tractors we needed to buy, by size, for the next twenty years. Along with the cost of buying new tractors and selling used-up tractors, I had to include operating expenses, including overhauls for all tractors. Once all the savanna had been converted to pasture, we could sell all the dozers and tractors that had been used to clear the land. I loved this work. I developed some very detailed spreadsheets.

The problem with this job was my commuting time. It took me twenty minutes to walk to the subway, and I spent another twenty-five minutes on the subway before I had to catch a bus. I spent thirty minutes on the bus and another fifteen minutes walking from the bus to the building. Once, when it was raining heavily, it took me more than three hours to return home. Each day I spent a minimum of three-plus hours commuting. Because my boss was a driven man, it was not possible to spend fewer than ten hours at work. I was tired all the time and did not have much of a family life. And I had to wear a suit every day.

I liked working at the bank, however. My boss was in charge, and he actively supervised. Everyone knew what was expected and was doing it. There was no gossiping allowed and no backstabbing. Everyone worked hard and was nice to each other. The boss set the tone. The difference between working here and in Natal was night and day. Yes, I loved Banco Safra.

One day our boss told us that we had to fly to the ranch and check its inventory of parts. I liked the idea of visiting the ranch, but not the idea of checking inventory. We flew from São Paulo to Brasilia on Sunday. From Brasilia we chartered a two-prop plane to fly us to the ranch, since it had its own landing strip

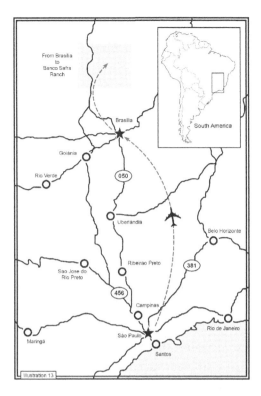

There were four of us traveling: the grass man, the cow man, the boss man, and me. A Jeep met us and took us to a large guesthouse, where we had dinner and were shown to our rooms. The accommodations were excellent, as was the cook.

On Monday morning we had breakfast early and set off to the ranch headquarters, where we met the project manager. He never seemed to be in a good mood, although being responsible for a project whose annual budget was more than ten million dollars could take the smile from a person's face.

I did not understand what we had to do with the inventory. The ranch had so many items in inventory, especially if we took into account tractor parts, fence-building materials, and all the other things—were we really going to count all that stuff? The project manager and our boss engaged in some straight talk for nearly an hour, and then we no longer needed to count anything. I had no idea what had just happened or why four of us had traveled here, except that it was important for us to see the actual ranch for which we were doing all the planning.

We drove around and saw each stage of work, including the clotheslining of the trees, the windrowing, the bunching, the burning, and so on. The fences were new, and the cattle were gigantic. They were all Zebu breed, huge animals with a hump on their shoulders, and they were ill-tempered.

During a group meeting, our grass man said he thought that the grass could be established quicker if the seed were protected by thin mulch. The project manager declared that they would plant two thousand acres of rye as an experiment. His people put in an order for two John Deere combines. They would sell the rye seed and then bale the straw, which would be used eventually as mulch. If that worked, they would expand the enterprise.

After lunch we went to participate in vaccinating cows and weaning calves. When around their calves, Zebu cows were more than ill-tempered—they were nasty and dangerous. Thus, the ranch had wooden corrals that were at least six feet high. The ranch workers drove the cows and calves through a narrow but tall chute. On the outside of the chute, there was a ledge about five feet from the ground. We stood on this ledge with our knees braced against the top fence board and with syringes in our hands. We each had a different vaccination. As the cow came through the chute at forty miles an hour, we leaned forward and gave the cow a dose of whatever vaccination we had been assigned. The calf always followed at the cow's heels. Ahead of me a man controlled a gate. He sent the cow to the left and the calf to the right. As soon as the cow discovered this, she went berserk. She cried, mooed, threw dirt with her front feet, and complained for all to hear.

Behind us were several empty pens. Once I turned around from my task and saw a mad cow pawing the dirt, her eyes focused on my backside. I yelled to the others, and we all jumped to the other side of the chute. She attacked the spot where I had been. She was an unhappy mama and had managed to jump two six-foot fences to reach a spot where she envisioned taking her revenge on us. Cowboys came and took her back to the pen she was supposed to be in.

There was time for us to talk to people who worked on the ranch. One man told us of the river that ran through the land. It was a huge river. He said that people there had caught fish as large as a man. He said that the ranch had trouble with the Indians coming up the river and stealing the ranch's boats. A couple Indians had walked into the bank's home base, put a canoe on their heads, and walked out. They had done this during the day with no attempt to hide what they were doing. They had bows and arrows slung on their shoulders. The ranch just marked this down as an expense. The company did not want to start a war with the Indians.

Inflation continued increasing a little each month. After I had worked at Banco Safra a few months, inflation had reached 30 to 35 percent a month and was still increasing. The bank had to do something to keep its employees from revolting. It gave automatic salary increases of 22 percent per month for two months, and on the third month, it gave a double increase, which amounted to a 49 percent increase. That combination of salary increases averaged out to be 30 percent per month. That helped, but my problem was that my rent payment for my house would be adjusted for inflation in December. My best estimate was that it would be ten times greater. If I paid 100 dollars per month now, it would increase to 1,000 dollars a month. I did not see how I could afford that. My only solution was to move farther from work, and that was not a viable option as far as I was concerned. We were at an impasse.

Katia and I had many conversations about our future. Kevin was going to school and being taught in Portuguese. Nick was going to preschool, and Christianne was four years old. Finally, we decided that we would return to the US, after living eight years in Brazil. For the most part, the children could start school in the US.

PREPARING TO RETURN TO THE US

I learned that to obtain my permanent exit visa, I needed about twenty documents, and they all had to be obtained during the last thirty days before I left Brazil—and this was for each state that I had lived in. I provided a copy of the list to Seu José, and he and his friends told me not to worry.

I started in São Paulo. First, I found that to obtain one document, I often needed to acquire another three or four documents, and the relevant offices were scattered all over the city. Sometimes, to obtain one document I had to speak with a specific person who happened to be on a two-week vacation. No one else dared give me the document—the man on vacation was the boss, and it was his and only his responsibility. I always had to pay for a document, but occasionally, an office did not take money at the point where the document was issued. I would have to go to a bank across town to pay. It was never the bank next door. It was never simple, in fact; it was always complicated.

It took me forever to obtain the documents. I was afraid their validity would start to expire before I got all the necessary papers. Sometimes I just wanted to sit down in a quiet place and cry.

We were also selling furniture, cars, and other belongings. As I received *cruzeiros* (the Brazilian currency at that time), I hurried to change them into dollars before inflation could devalue them any further. To do this, I had to go to the black market since it was very complicated, or even impossible, to make this exchange at the Banco do Brasil, the only bank where I could legally exchange *cruzeiros* for dollars. I found an illegal black-market exchange ten feet in front of the national police office. That was why I loved Brazil. Laws meant nothing. That was the *jeito* concept at work. There was always a way around the law, and sometimes it was only a few feet in front of the police station.

Finally, we had sold everything we could and had separated what we were taking with us. I had my documents, and we had exchanged all our money for US dollars. Then I learned that Katia needed a visa to reenter the US, and it would take three months to obtain it. Well, that was not in the plans. So I had the moving company pick up our things to send them to my sister's place in the US. On the day I was to leave Brazil, I loaded Katia and the children onto a plane to Natal in the morning. They would spend time with her family before moving on to the US. This was actually as it should have been. Dona Naide and Seu José deserved time with their daughter and grandchildren before an extended stay in the US.

I spent the day waiting for my evening flight. I watched two movies to help me pass the time. Eventually, my flight departure time arrived. It was August 1, 1987. When the plane accelerated for takeoff, my mind was filled with excitement and terror. I had no job, very few possessions, very little money, and a family depending on me. I was starting over from scratch. My sister Shelli and her family were collecting me at the airport, and I knew that I could spend a few days with her in her house. After that, everything was unknown.

I had a long layover in Miami, during which time I was frantic. I had been out of the country for eight years. I wondered how much had changed. I also could not formulate sentences in English. Invariably, I would insert Portuguese words and use Portuguese sentence structure. It was not pretty. After what seemed like an eternity, my flight left for Atlanta and eventually Omaha.

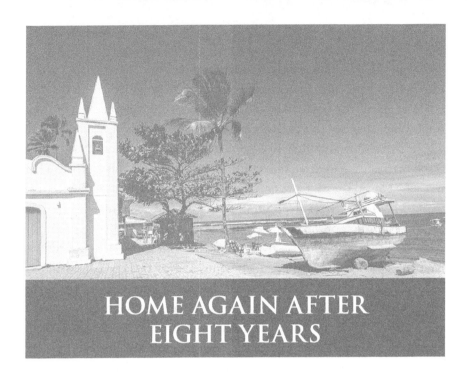

HOME AGAIN AFTER EIGHT YEARS

I was speechless when I saw my little sister Shelli and her family. They piled my luggage into the car trunk, and we were off. Everyone wanted to speak with me, but I was so overwhelmed by everything that I could not speak back. Sometimes conversation was uncomfortable because it took me so long to answer their questions. And the children had grown so much.

Halfway home, they wanted to stop at a steakhouse and eat. I was hungry, but I was still overwhelmed and could not eat. They all ordered food, and I ordered a beer, followed by another beer. By the time I started to relax, it was already time to leave. I had eaten nothing and now was regretting that fact because I was hungry.

I slept like a log that night—both from the relief of being back in the US and because of the beers that I had drunk. In the morning I was very hungry, and I ate well. Shelli and her husband Bob agreed to rent me the house where Bob had grown up. I could pay a low rent. The house was located at the end of two roads and next to the Loup River. That was perfect. I wanted wilderness. There would not be any traffic passing our house. One road that dead-ended by our house had been abandoned for decades and could be traveled only by four-wheeled drive vehicles. The other dead-ended at our driveway.

FINALLY, A JOB

I bought an old beat-up car and started looking for a job. After a few day jobs such as dishwashing, I got a job in Columbus, Nebraska, in a manufacturing plant. I would travel forty miles in each direction, half this distance on gravel or dirt roads. My job was to operate a robot welder, so I stood on my feet for eight hours and pushed buttons. The company I worked for manufactured seats for Ford cars and vans and sent the seats to the company's factory in Kansas City.

After a couple of weeks, we went to an obligatory nine hours a day, and then a couple of weeks later, we moved to ten hours a day. My checks were improving. I made friends with my supervisor and told him I needed to find a way to make more money. He told me that there was an opening for lead man in the shipping department. He recommended me, and I got the job and a small raise. Lead man was the number-two man in shipping behind the manager, who was on salary.

It did not take long before we also had to work Saturdays. Sometimes in the shipping department I had to work twelve hours for each of the six days. My paycheck was beginning to impress me. However, the manager was not happy because he was salaried and received the same money whether he worked forty hours a week or eighty hours a week. I was making twice what he did—not fair for him, but great for me. The only problem was that my feet bothered me from all the time I spent walking on concrete. My hands also hurt from having to move and manipulate the metal pieces that were manufactured.

When I awoke in the mornings, I was exhausted, and I could not move my feet or my hands. I had to use one hand to massage the other until the stiffness and pain subsided. Then I massaged my feet. I had time to eat breakfast, and then I hit the road for another twelve-hour shift.

On October 8, 1987, Katia and the children arrived from Natal. I was so happy to see them. I did not want to be alone anymore. The children loved the farm. There were many old buildings, straw piles, and pigs and cows, and they never got bored. Once Nicholas climbed onto the roof of a small shed, and when he jumped toward the ground, his belt snagged a nail. Kevin kept playing, but Nicky's yells finally attracted Katia's attention. She had to call Uncle Bob, who rushed to save Nicky from the roof. In fact, Uncle Bob was always saving Nick or Katia or someone.

Katia was not so sure about living on a farm with hogs and cattle. Once while cleaning the bedroom, she looked out the window and saw a cow looking back at her. She called Uncle Bob for help. Katia explained that there was an angry cow looking at her through the window. Bob came and guided the old cow back into the corral.

Katia wanted a car to be able to take the kids to school, so I bought an old four-speed Toyota pickup for me and gave her the ugly car I'd bought upon returning. The pickup was old and ugly too, but I loved my Toyota. It took me to work and brought me home at 3:00 a.m. always. In the winter during a blizzard, the twenty miles of old country roads were drifting badly. I put her in third gear and stepped on the gas. Each snowdrift tried to grab us, but after a split second of doubt, my Toyota pounded through the drifts and took me home safely.

We had one small problem. Kevin and Nicholas were both school-age, and their school was a two-room country school in a Polish community. Kevin and Nicholas did not speak English, not a word. After getting home at 3:00 a.m., I had to get up and take them to school and sit with them for several hours, acting as a translator. Then I went to my twelve-hour shift. For three or four weeks, I thought I would die. I just wanted to sleep, but Kevin and Nicholas picked up English quickly. The other children and teachers were helpful.

At Thanksgiving the company gave all employees a five-pound ham.

At break time we were instructed to take them to our vehicles and lock our doors so that no thefts could occur. When I reached my pickup and was about to place my ham on the car seat, I saw my friend at his car. I yelled, "Hey, Gus! Isn't it great they gave us all a ten-pound ham?" To my surprise, he answered that it was. I looked around, and men and women were stopping in their tracks and returning to their vehicles. Others were holding up their five-pound ham toward the parking lot safety lights. Still others had their cigarette lighter out. I heard several exclamations of "Oh shit!" followed by a door slamming shut. I smiled and headed back inside the break room. It was so easy.

Our arrangement with the Kansas City Ford plant was that we had to supply the plant with parts in the morning that it would need in the afternoon. It was a just-in-time agreement. If we neglected to do this and the Ford plant had to cease operations for lack of parts, our company had to pay a fine. I think it was more than fifty thousand dollars per hour. This was a situation our company wanted to avoid. Well, one day we had a situation. We had a truckload of car seats already headed toward Kansas City, but they were not going to arrive in time to keep us "just in time," and the fine would be applied.

In a desperate attempt to avoid the stain of the fine being applied on his watch, our plant manager ordered us to gather fifty or so such seats that we had lying around. If we got them to the airport in time, workers could tear out the seats in the company's plane, fill the plane with these seats, and fly them lickety-split to Kansas City. I supposed the manager intended for the plane to land at the Ford plant and save the day.

The manager asked for volunteers. One guy volunteered his pickup, and the plant manager found a trailer to haul the seats and I volunteered to accompany him. The plant manager was so hyper that he made everyone he saw nervous. Several people piled the seats into the trailer in a haphazard way. The plant manager patted the pickup driver on his shoulder and told him to hurry. He did. He pulled into traffic and put his pedal to the metal. In fact, he was driving at an unsafe speed with a trailer, which started to swing back and forth across the lane. I felt the trailer pull on the pickup, and then the trailer flipped, throwing seats ahead of us into four lanes of traffic, two of them oncoming lanes. The people in oncoming cars saw these black specks sailing through the air, getting bigger and bigger until they crashed down in front of them. The drivers smashed on their brakes to avoid colliding with the flying seats and started gasping for air when they realized how close they had come to a serious accident.

No other cars or drivers were hurt in the accident, but it could have been so much worse. I was angry at the plant manager for winding up the pickup driver, who had just wanted to help the manufacturing plant. He could have been responsible for several deaths.

We immediately began working to right the overturned trailer and slow traffic down and pick up the damaged car seats. Every car seat was dented or scraped or bent. All we could do was return to the factory and tell the plant manager that his plan had not worked. His face was red, and I thought it was going to explode.

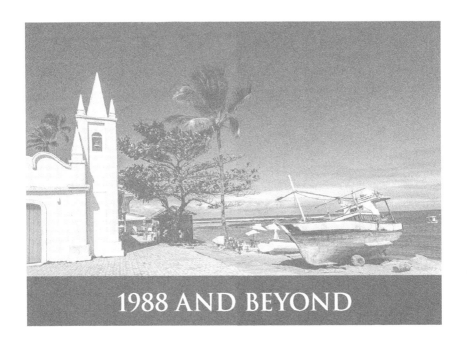

1988 AND BEYOND

I found a job as a farm manager with Farmers National Company, where I worked until mid-1995. I then moved to the St. Louis area and worked with an unscrupulous company that helped farmers market their commodities. After that I worked in a couple of unscrupulous market research companies. Tired of unscrupulous companies, in 1999 I started teaching statistics in the School of Business at Southern Illinois University at Edwardsville, where I taught until August 2015 when I retired in August, 2015.